HYPERREALITY

What comes after the Internet? Imagine a world where it is difficult to tell if the person standing next to you is real or a virtual reality and whether they have human intelligence or artificial intelligence; a world where people can appear to be anything they want to be. HyperReality makes this possible.

HyperReality: Paradigm for the Third Millennium offers a window into the world of the future, an interface between the natural and artificial. Nobuyoshi Terashima led the team that developed the prototype for HyperReality at Japan's ATR laboratories. John Tiffin studied the way HyperReality would create a new communications paradigm. Together with a stellar list of contributors from around the globe who are engaged in researching different aspects of HyperReality, they offer the first account of this extraordinary technology and its implications.

This fascinating book explores the defining features of HyperReality: what it is, how it works and how it could become to the information society what mass media was to the industrial society. It describes ongoing research into areas such as the design of virtual worlds and virtual humans, and the role of intelligent agents. It looks at applications and ways in which HyperReality may impact on fields such as translation, medicine, education, entertainment and leisure. What are its implications for lifestyles and work, for women and the elderly? Will we grow to prefer the virtual worlds we create to the physical world we adapt to?

HyperReality at the beginning of the third millennium is like steam power at the beginning of the nineteenth century and radio at the start of the twentieth century, an idea that has been shown to work but has yet to be applied. This book is for anyone concerned about the future and the effects of technology on our lives.

Contributors: Narendra Ahuja, Prem K. Kalra, Nadia Magnenat-Thalmann, Laurent Moccozet, Minako O'Hagan, Lalita Rajasingham, Katsunori Shimohara, Sanghoon Sull, Nobuyoshi Terashima, John Tiffin.

Editors: John Tiffin is Professor Emeritus of Communication Studies at Victoria University of Wellington and Chancellor and founder of the Global Virtual University. **Nobuyoshi Terashima** is the Dean of the Graduate School of Global Information and Telecommunication Studies at Waseda University, Japan.

HYPERREALITY

Paradigm for the Third Millennium

Edited by
John Tiffin and Nobuyoshi Terashima

London and New York

First published 2001
by Routledge
11 New Fetter Lane, London EC4P 4EE

Simultaneously published in the USA and Canada
by Routledge
29 West 35th Street, New York, NY 10001

Routledge is an imprint of the Taylor & Francis Group

Selection and Editorial Matter © 2001 John Tiffin
and Nobuyoshi Terashima; individual chapters, respective authors

Typeset in Garamond by Florence Production Ltd, Stoodleigh, Devon
Printed and bound in Great Britain by Biddles Ltd, Guildford and King's Lynn

British Library Cataloguing in Publication Data
A catalogue record for this book is available from the British Library

Library of Congress Cataloging in Publication Data
HyperReality : paradigm for the third millennium / edited by John Tiffin and
Nobuyoshi Terashima.
    p. cm.
Includes bibliographical references and index.
1. Reality. 2. Virtual reality. 3. Computers—Social aspects. I. Tiffin, John. II.
Terashima, Nobuyoshi.

BD331 .H97 2001   2001019768
306.4'6—dc21

ISBN 0–415–26103–1 (hbk)
ISBN 0–415–26104–X (pbk)

CONTENTS

FIGURES

CONTRIBUTORS

Narendra Ahuja
Dr Narendra Ahuja is a professor in the Department of Electrical and Computer Engineering, the Coordinated Science Laboratory and the Beckman Institute at the University of Illinois at Urbana-Champaign. His interests are in computer vision, robotics, image processing, image synthesis, sensors and parallel algorithms. His current research emphasises integrated use of multiple image sources of scene information to construct three-dimensional descriptions of scenes; the use of integrated image analysis for realistic image synthesis; parallel architectures and algorithms and special sensors for computer vision; and use of the results of image analysis for a variety of applications, including visual communication, image manipulation, video retrieval, robotics and scene navigation. He received the 1998 Technology Achievement Award of the International Society for Optical Engineering. He is co-author of *Pattern Models* and *Motion and Structure from Image Sequences*. He also co-edited *Advances in Image Understanding*.

Prem Kalra
Dr Prem Kalra is Associate Professor in the Department of Computer Science and Engineering, at the Indian Institute of Technology, New Delhi, India.

Nadia Magnenat-Thalmann
Professor Nadia Magnenat-Thalmann has pioneered research into virtual humans over the last 15 years, participating in and demonstrating some of the most spectacular state-of-the-art developments in the field. She is responsible for the rigorous and intensive academic research programs that made them possible. She was a professor at the University of Montreal in Canada from 1977 to 1988 and received several awards during this period, including the prestigious 1985 Communications Award from the Government of Quebec, and the 1987 Woman of the Year Nomination from the Montreal Urban Community (on public vote).

In 1989, on her return to Switzerland, she founded the MIRALab, an interdisciplinary creative research laboratory at the University of Geneva. She contributed to the creation of the Department of Information Systems, where she has been Director from 1996 to the present time. She was elected a Member of the Swiss Academy of Technical Sciences in 1997.

She has contributed to the publication of more than 200 scientific papers, and has directed and produced several films, among them *Dream Flight* (1982), *Rendezvous in Montreal* (1987), *Marilyn in Geneva* (1995), *The Terracotta Soldiers Army* (1997), *Marilyn at the United Nations* (1997), *Cyberdance* (1998) and *Fashion Dreams* (1998).

Laurent Moccozet

Dr Laurent Moccozet is a computer scientist working at the University of Geneva. His main areas of interest are geometric modelling, geometric deformations and their application to human modelling and animation. His research on hand modelling and animation for virtual humans was part of the HUMANOID 1 and 2 European projects in interaction in 3D modelling and deformation of the body while in motion.

Minako O'Hagan

Dr Minako O'Hagan, originally from Japan, has lived in New Zealand for the last 17 years, during which time she has worked in the public and private translation sectors as a Japanese/English translator and editor. After gaining an MA in Communications from Victoria University of Wellington in 1991 on the study of the impact of information technology on the translation industry, she developed her thesis into a publication *The Coming Industry of Teletranslation* (O'Hagan 1996), establishing a new area of translation studies. Her doctoral research, conducted at Waseda and Victoria Universities, focused on a new form of language support in HyperReality, which is the subject of her chapter. She is currently co-authoring a book on the emergence of teletranslation and tele-interpretation on the Internet.

Lalita Rajasingham

Dr Lalita Rajasingham is a Senior Lecturer at Victoria University of Wellington where she has acquired an international reputation for her research into the implications of new information technologies for education and for her concern for their cultural implications, especially in developing countries. With John Tiffin she has been involved in a long-term action research project to develop a virtual class and a virtual university. Beginning in 1986, this research led to a book that outlined the concept of the virtual class (Tiffin and Rajasingham 1995) and a sequel on the virtual university is planned.

Katsunori Shimohara

Dr Katsunori Shimohara is a leading expert in Japan on artificial life and evolutionary systems, and has been working on computer–human communications since 1993 as Head of the Evolutionary Systems Department at ATR Human Information Processing Research Laboratories. From 1999 to mid-2000 he held the concurrent post of Executive Manager, Social

Communication Laboratory, NTT Communication Science Laboratories. He has been Director of the Information Sciences Division of ATR International since mid-2000 and Guest Professor in the Graduate School of Informatics, Kyoto University since 1998. His publications include the 1998 book, *Artificial Life and Evolving Computers* (in Japanese).

Sanghoon Sull

Dr Sanghoon Sull is an Associate Professor in the School of Electrical Engineering, Korea University. He was a post-doctoral research associate at the Beckman Institute, University of Illinois, Urbana-Champaign during 1993–4. He went on to research video-based obstacle detection at NASA, Ames Research Center and later video indexing and browsing at IBM, Almaden Research Center. His current research interests include multimedia data management, including indexing and retrieval, image understanding and processing and computer graphics.

Nobuyoshi Terashima

Dr Nobuyoshi Terashima is currently Dean at the Graduate School of Global Information and Telecommunication Studies, Waseda University, Tokyo, Japan and was President of ATR Communications Systems Research Laboratories from 1991–96, where he led the team that developed the prototype on which the concept of HyperReality is based. Prior to that, he was involved in research into artificial intelligence at NTT (Nippon Telegraph and Telecommunications). Since 1996 he has been a professor at Waseda University where he teaches courses on HyperReality and where he has founded the Graduate School of Global Information and Telecommunication Studies. He also contributed to the formation of GITI (the Global Information and Telecommunication Institute) under whose aegis he has continued his research on HyperReality. Within this he has a special concern for the development of the HyperClass. Terashima is one of the pioneers of VR in Japan and a leading authority on artificial intelligence. His many books in Japanese have popularised the key ideas that are making the country an information society.

John Tiffin

John Tiffin is Professor Emeritus of Communication Studies at Victoria University of Wellington and Chancellor and founder of the Global Virtual University. The latter is an initiative by academics from around the world to establish on the Internet a virtual university that offers global skills relating to the future. It is hoped that, in due course, it will be the first HyperUniversity. From 1985 to 1998 he was the David Beattie Professor of Communications at Victoria University, in which role he founded communication studies in New Zealand. He has been a headmaster in Ethiopia and a television producer in Brazil, a company director in Washington DC and manager of the research division of a software company in New Zealand.

Throughout all these activities he has had a passion for the future of education.

In 1995 he published, with Dr Lalita Rajasingham, *In Search of the Virtual Class: Education in an Information Society*. This was also published in Spanish and Russian.

PREFACE

John Tiffin

In 1992 I convened a meeting in New Zealand of people interested in commercially designing virtual realities (VR). We met to discuss whether the Head Mounted Display Unit (HMDU) and gloves type of VR then becoming commercially available constituted a technological platform that warranted applications development. VR parlours with games like Pterodactyl were beginning to appear. Were we at the stage that film was at when it was projected in cinemas or were we at the earlier transitory stage of kinetoscope peep-show parlours?

While we were arguing the matter a fax broke into life in the corner of the conference room and out of the blue came an invitation from Nobuyoshi Terashima to look at a new VR technology he had developed that used telecommunications. It was as though he had been a telepresence at our meeting and had provided a tele-answer to our question.

Dr Terashima was at that time President of ATR Communications Systems Research Laboratories. The ATR (Advanced Telecommunications Research) building is located in Kansai Science City and is Japan's powerhouse for basic research into advanced telecommunications technologies. Terashima was leading a 10-year project (1986–96) that was to result in the prototype for HyperReality.

At the beginning the project was focused on developing technology for teleconferencing in virtual reality, but by 1992 Terashima had begun to look beyond this to the implications of having real people and real objects commingle with virtual people and virtual objects in a relatively seamless way. What would it mean for manufacturing, for medicine, for the way business was conducted? It was to explore such questions that he asked me to visit ATR and talk about the work I was doing with Dr Lalita Rajasingham on the 'Virtual Class' (Tiffin and Rajasingham 1995).

In industrial societies it is transport systems and buildings that bring together the critical components of education: teachers, students, knowledge and practice. The device that brings the components of a conventional educational system into conjunction is called a classroom. Rajasingham and I had been studying the possibilities of using telecommunications to bring together the critical components of education. We called this a 'Virtual Class'. It would mean that, in an information society, access to education would no longer depend on where people lived.

Since 1986 Rajasingham and I had been experimenting with forms of teleconferencing for instruction. However, telecommunications technology in 1992 was still too primitive for it to be any more successful at tele-education than predecessors such as educational television. There had to be a better way. When I visited ATR laboratories I realised that there was. Terashima had a technology looking for applications. We had an application looking for a technology.

So began a series of discussions, which in due course led to this book. Terashima and I were seeking to understand the nature of what, by 1995, Terashima had begun to call HyperReality (HR). By this he meant a technological infrastructure that supported the seamless interaction of virtual people and objects with real people and objects and human intelligence with artificial intelligence. Debate took place in restaurants that ranged from Waikiki to Wellington and Kyoto to Canberra and was enriched by the fact that, as well as being one of Japan's leading computer scientists, Terashima is also a distinguished gourmet.

Initially we talked of what will be possible with commercial versions of the prototype technology developed in the ATR laboratories. Then we began to think about what could follow as the principles of HyperReality become embodied in the sophisticated technological systems we can reasonably envisage for the next century. As we did so, we began to realise that what we were talking about was the technological infrastructure for communications in the post-industrial society.

REFERENCE

Tiffin, J. and Rajasingham, L. (1995) *In Search of the Virtual Class*, London: Routledge.

ACKNOWLEDGEMENTS

Dr Margaret Allan of James Cook University has had the role of editors' editor and as such has profoundly influenced the shape this text takes. In a book with such a distribution of subject and style, where for most of the writers English is a second language, she has been of enormous help in preserving continuity. Whatever synergy is achieved is hers, whatever anarchy remains is ours.

The insight in the illustrations for the chapters by Dr Terashima and myself comes from the fact that the artist Koji Matsukawa is also working on the design of virtual realities for HyperReality.

The book would not have been possible without the award of a research fellowship by the Telecommunications Advancement Organisation (TAO) of Japan, which allowed me to work with Dr Terashima and the HyperClass project. I would like to express my gratitude to TAO and to my hosts at Waseda University in the Global Information and Telecommunication Institute (GITI).

Finally I would like to acknowledge the role of Victoria University of Wellington in supporting my work on HyperReality during my time there as the David Beattie Professor.

INTRODUCTION

John Tiffin

HyperReality (HR) is a hypothetical communications infrastructure made possible by information technology. It allows the commingling of physical reality with virtual reality and human intelligence with artificial intelligence. Something like HyperReality is going to emerge in the first century of the new millennium. It may not be called HyperReality and it may not take the precise forms described in this book, but the essence of what we call HR will be there and it will change the world. It will impact on the industrial society like mass media impacted on the *gemeinschaft* of pre-industrial societies. It is what the Internet will become and the place where computers, telecommunications, artificial intelligence and virtual reality are taking us.

Durand (1981) sees the future as a trajectory that is as yet untraced through a web of possible directions. These directions are worthy of scientific study even though the object of study is by its nature virtual: 'the future does not exist'. The book can, therefore, be thought of as a research essay in designing the future, which seeks to explore a number of possibilities suggested by our present technological knowledge. The variety of exploration ranges from research into the specifics of the technology to the forecasting of applications using the techniques of futures scenarios.

The book does not seek to fit within the dialectic of modernists and postmodernists who see the information society as a place that, like Milton's mind, makes 'a heaven of hell and a hell of heaven' (*Paradise Lost* 225). Rather it aims for the mode of Drexler's (1990) *Engines of Creation*, which gave impetus to the study of nanotechnology by arguing the possibility of having machines and computers at the level of molecules and then describing the consequences for society of realising that possibility.

Like other original thinkers, Terashima has the gift of reducing a powerful idea to its essence, where it can be expressed in simple formulas that then provide the basis for technological development. He does this in Chapter 1 where he defines HyperReality, provides a concept structure for its components and discriminates it from such adjacent technologies as augmented reality and mixed reality.

1

Chapter 2 seeks to place HyperReality in perspective for someone like myself with a background in communications rather than information technology. My main concern is to understand its possibilities and limits as a new medium and I approach this by viewing HR as a communications paradigm. Print, radio, television and film were the mass media that shaped the industrial society. How will HR shape the information society?

These first two chapters bring to mind the modus of Shannon and Weaver's *The Mathematical Theory of Communication* (1948). In this book Claude Shannon expounded his theory of communications with a mathematical rigour that has stood the test of fifty years and established the body of theory behind information technology. In an extended foreword, Warren Weaver sought to explain the sociological implications of Shannon's theory and how it could be applied to all communications. Shannon was addressing people with a background in science and technology. Weaver was addressing people in the social sciences and the emerging discipline of communications studies. They were looking at the same coin but from different sides.

On Shannon's side of the coin could have been written: 'for people studying information' and on Weaver's side: 'for people studying communications'. Half a century later 'communicators' and 'information technologists' still view the same phenomenon from opposing perspectives. They are the inheritors of C. P. Snow's (1959) 'two cultures'. One of the frightening factors of our time is that those who are building the information society seem unable to develop a meaningful discourse with those who will live in it.

This book seeks to speak to both cultures. It is for people in the information sciences and people in the communication arts who seek to understand and effect the future. The book is in fact divided into a section on the technology and a section on applications. However, it is hoped that there will be some crossover between the readers from the two cultures. Chapters 1, 2 and 9 have both kinds of readers in mind and each chapter has an introduction that allows the reader to judge its relevance.

Shannon's theory explains how people can communicate by telecommunications using written and spoken words and 2D images. In Chapter 3 Narendra Ahuja and Sanghoon Sull explain how to communicate by telecommunications using 3D images. They address the fundamental issue in HyperReality of making it possible for participants to see immersive three-dimensional images derived from cameras in another location. The authors describe their research in 3D computer vision which led them to develop a technique called analysis-guided video synthesis. This makes it possible to extract 3D images from the 2D images of conventional still and video cameras, such as the small golfball-sized video cameras that are being used more and more for surveillance and for videoconferencing on the Internet.

HyperReality is where physically real people meet virtual people who may, using the techniques described by Ahuja and Sull, be projections of real people. Alternatively they may be generated using a computer. In

2

Chapter 4 Nadia Magnenat-Thalmann, Prem Kalra and Laurent Moccozet describe their work in generating autonomous virtual humans.

HyperReality is where human and artificial intelligence interact. In Chapter 5 Katsunori Shimohara looks at the growing role of intelligent agents and their implications for HyperReality.

Part 2 of the book looks at applications of HyperReality. In Chapter 6 Minako O'Hagan extends her research on teletranslation to consider the implications of people from different cultures with different languages meeting in HyperReality and the attendant need for 'HyperTranslation'. In Chapter 7 Lalita Rajasingham and I examine the implications of Hyper-Reality for education and describe the research we have been involved in with Terashima to develop a HyperClass. In Chapter 8 I examine the many aspects of leisure in HyperReality and in doing so try to paint a picture of life in HyperReality.

The last chapter seeks to place the development of HyperReality in a futures scenario. In writing it Terashima and I came to realise that, besides juxtaposing science-based information culture with arts-based communication culture, we were also trying to make the 'twain' of East and West meet. I approached the subject from an essentially British background and thought about it in English while Terashima's background and language are Japanese. There is a similar East/West balance between the other authors. I have personally found the pursuit of an understanding of what Hyper-Reality means within this opposition of cultures at once the most fruitful and the most frustrating aspect of editing and contributing to the English version of this book.

One of the great pleasures of our time, which hopefully HR will extend, is the globalisation of collegiality. Academic collegiality normally comes about through a shared interest in a subject. Yet the authors in this case are all established in very different fields: communications, education, tele-translation, computer science and artificial intelligence. It is Terashima with his concept of HyperReality who provides a common ground where each of us can find a place for our ideas. Tentatively the components of this book are those of a new field of study.

REFERENCES

Drexler, E.K. (1990) *Engines of Creation*, London: Fourth Estate.
Durand, J. (1981) *Les Formes de la Communication*, Paris: Editions Dunod.
Shannon, C. and Weaver, W. (1948) *The Mathematical Theory of Communication*, Urbana: University of Illinois Press.
Snow, C.P. (1959) *The Two Cultures and the Scientific Revolution*, Cambridge: Cambridge University Press.

1

THE DEFINITION OF HYPERREALITY

Nobuyoshi Terashima

Editors' introduction
HyperReality (HR) is a technological capability like nanotechnology, human cloning and artificial intelligence. Like them, it does not as yet exist in the sense of being clearly demonstrable and publicly available. Like them, it is maturing in labora-tories where the question 'if?' has been replaced by the question 'when?' And like them, the implications of its appearance as a basic infrastructure technology are profound and merit careful consideration.

In this chapter Nobuyoshi Terashima defines HR. He does this by specifying the elements involved and their relationships and by discriminating HR from associated technologies. This is a schematic description of HR for those seeking the technolog-ical roots of the subject. Terashima then proceeds to illustrate the definition with examples of HR and to survey some of its possible applications. Finally, he exam-ines some of the key technologies that will need to develop for the use of HR to become common practice.

INTRODUCTION

The concept of HyperReality (HR), like the concepts of nanotechnology, cloning and artificial intelligence, is in principle very simple. It is nothing more than the technological capability to intermix virtual reality (VR) with physical reality (PR) and artificial intelligence (AI) with human intelligence (HI) in a way that appears seamless and allows interaction.

The interaction of HI and AI is a developing function of communica-tions and telecommunications. The interaction of PR and VR in HR is made possible by the fact that, using computers and telecommunications, images from one place can be reproduced in 3D virtual reality at another place. The 3D images can then be part of a physically real setting in such a way that physically real things can interact synchronously with virtually real things. It allows people not present at an actual activity to observe and engage in the activity as though they were actually present. The technology

4

will offer the experience of being in a place without having to physically go there. Real and unreal objects will be placed in the same 'space' to create an environment called a HyperWorld (HW). Here, imaginary, real and artificial life forms and imaginary, real and artificial objects and settings can come together from different locations via information superhighways, in a common plane of activity called a coaction field (CF), where real and virtual life forms can work and interact together.

Communication in a CF will be by words and gestures and, in time, by touch and body actions. What holds a coaction field together is the domain knowledge (DK) that is available to participants to carry out a common task in the field. The construction of infrastructure systems based on this new concept means that people will find themselves living in a new kind of environment and experiencing the world in a new way.

HyperReality is hypothetical. Its realisation as an infrastructure technology is in the future. Today parts of it have a half-life in laboratories around the world. The experiments that demonstrate its technical feasibility depend upon high-end silicon graphic workstations and assume broadband telecommunications. These are not yet everyday technologies. HR is based on the assumption that the technological trends on which it is based will continue (see Figure 1.1), that Moore's law will operate, that computers will get faster and more powerful and digital information superhighways will provide megabandwidth with and without wires. Nanotechnology makes feasible the idea of wearable computing. Voice recognition, image recognition, gesture recognition and writing recognition are developments that lead toward the barrier-free interfaces that are at the heart of HR. VRML (Virtual Reality Modulating Language) chat worlds such as Community Place and Active Worlds give a first idea of what a HyperWorld could look like. Distributed VR, Augmented VR and Mixed VR are conceptual steps towards HR. Machine Learning, Navigation Agents and the developing association of intelligent agents and avatars are movements in the direction of AI in HR. All that is lacking is the kind of integrating vision that this book seeks to supply.

The project that led to the concept of HR began with the idea of teleconferencing in virtual reality. It was the theme of one of the first four labs at ATR (Advanced Telecommunications Research) in Kansai Science City. Likened to the Media Lab at MIT or the Santa Fe Institute, ATR has acquired international recognition as Japan's premier research centre concerned with the telecommunication and computer underpinnings of an information society. The research lasted from 1986 to 1996 and successfully demonstrated that it was possible to sit down at a table and engage interactively with the telepresences of people who were not physically present (Figure 1.2). True, their virtual personas, which we now refer to as avatars, looked like tailor's dummies and moved jerkily. However, it was possible to recognise who they were and what they were doing and it was possible for real and virtual

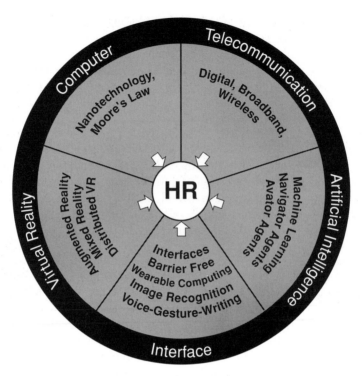

Figure 1.1 Technology trends contributing to the entitation of HyperReality.

people to work together on tasks that required manipulating virtual objects (Terashima *et al.* 1995; Terashima 1993; 1994a; 1994b; 1995b).

The technology involved comprised a large screen, a camera, data gloves and glasses. Virtual versions were made of the people, objects and settings involved and these were downloaded to computers at different sites. Then it was only necessary to transmit changes in position, shape and movement in addition to sound. As long as one was orientated toward the screen and close enough not to be aware of its edges, interrelating with the avatars appeared seamless. Was it possible to extend the screens? Why not a room where some of the walls and parts of the ceiling and floor were, in effect, screens? A mirrored wall can give the illusion of an extension to a room. Looking at oneself and others in the mirror can give an eerie feeling of seeing other people in an extended place. Could careful matching of virtual rooms on to the walls of real rooms give a similar effect?

The effect of mixing virtual reality and physical reality can also be generated by using glasses and gloves. This requires glasses with a dual function. First, the function glasses normally have of adjusting light reflected from objects in the wearer's line of vision, to make what they are looking at clear. In other words to help people see physical reality. Second, the func-

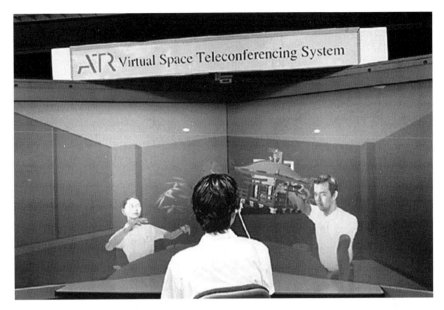

Figure 1.2 Experiment at ATR: Real and virtual people interacting in the construction of a virtual model of a shrine.

tion of projecting images of virtual objects derived from a computer. In other words to show virtual reality. The trick then is to interlace the virtual and real images so that they make sense together. There are head-mounted display (HMD) systems that do this (Bajura, Fuchs and Ohbuchi 1992).

Most humans understand their surroundings primarily through their senses of sight, sound and touch. Smell and even taste are sometimes critical too. As well as the visual components of physical and virtual reality, HyperReality needs to include associated sound, touch, smell and, although the idea is on the back burner for the moment, even taste. The technical challenge of HyperReality is to make physical and virtual reality appear to the full human sensory apparatus to intermix seamlessly. It is not dissimilar to, or disassociated from, the challenges that face nanotechnology at the molecular level, cloning at the human level and artificial intelligence at the level of human intelligence. Advanced forms of HR will be dependent on extreme miniaturisation of computers. HR involves cloning, except that the clones are made of bits of information. Finally, and perhaps the most important aspect of HR, it provides a place for human and artificial intelligences to interact seamlessly. The virtual people and objects in HR are computer generated and can be made intelligent by human operation or they can be activated by artificial intelligence. Communication in HR could become an endless Turing test.

HYPERREALITY CONCEPT SCHEMATA

Definition of HyperReality (HR)

HyperReality (Terashima 1995a; Terashima and Tiffin 1999) is a techno-logical capability that makes possible the seamless integration of physical reality and virtual reality, human intelligence and artificial intelligence: HR = the seamless integration of (PR, VR, HI, AI) where PR = Physical Reality, VR = Virtual Reality, HI = Human Intelligence and AI = Artificial Intelligence.

HR makes it possible for the physically real inhabitants of one place to purposively coact with the inhabitants of remote locations as well as with computer-generated imaginary or artificial life forms in a HyperWorld.

A HyperWorld is an advanced form of reality where real-world images are systematically integrated with 3D images derived from reality or created by computer graphics. The field of interaction of the real and virtual inhab-itants of a HyperWorld is referred to as a coaction field (CF).

Definition of HyperWorld (HW)

HW is a seamless intermixture of a (physically) real world (RW) and a virtual world (VW). HW can, therefore, be defined as (RW, VW).

A real world consists of real natural features, real buildings and real objects. It is whatever is atomically present in a setting and is described as (SE), i.e. the scene exists.

A virtual world is whatever is present in a setting as bits of computer-generated information. It consists of the following:

- *SCA (Scene shot by camera)*: Natural features, buildings and objects that can be shot with cameras (video and/or still), transmitted by telecom-munications and displayed in VR.
- *SCV (Scene recognised by computer vision)*: Natural features, buildings, objects and inhabitants whose 3D images are already in a database and are recognised by computer vision, transmitted by telecommunications and reproduced by computer graphics and displayed in VR.
- *SCG (Scene generated by computer graphics)*: 3D objects created by computer graphics, transmitted by telecommunications and displayed in VR.

SCA and SCV refer to VR derived from referents in the real world where-as SCG refers to VR that is imaginary. A VW is, therefore, described as: (SCA, SCG, SCV). This is to focus on the visual aspect of a HyperWorld. In parallel, as in the real world, there are virtual auditory, haptic and olfac-tory stimuli derived either from real-world referents or generated by computer.

8

Definition of a coaction field

A coaction field is defined within the context of a HyperWorld. It provides a common site for objects and inhabitants derived from PR and VR and serves as a workplace or an activity area within which they interact. The coaction field provides the means of communication for its inhabitants to interact in such joint activities as designing buildings or playing games. The means of communication include words, gestures, body orientation and movement, and in due course will include touch. Sounds that provide feedback in performing tasks, such as a reassuring click as elements of a puzzle lock into place or as a bat hits a ball, will also be included.

The behaviour of objects in a coaction field conforms to physical, chemical and/or biological laws or to laws invented by humans. For a particular kind of activity to take place between the real and virtual inhabitants of a coaction field, it is assumed that there is a domain of knowledge based on the purpose of the coaction field and that it is shared by the inhabitants.

Independent coaction fields can be merged to form a new coaction field, termed the outer CF. For this to happen requires an exchange of domain knowledge between the original CFs, termed the inner CFs. The inner CFs can revert to their original forms after interacting in an outer CF. So, for example, a coaction field for playing cards could merge with a coaction field for bedside nursing to form an outer coaction field that allowed a nurse to play cards with a patient. The patient becoming tired, the CF for card-playing would terminate and the outer CF would revert to the bedside nursing CF.

A coaction field can therefore be defined as: CF = {field, inhabitants $(n > 1)$, means of communication, knowledge domain, laws, controls}.

In this definition a field is the locus of the interaction that is the purpose of the coaction field. This may be well defined and fixed as in the football field of a CF for playing football or the tennis court of a CF for playing tennis. Alternatively, it may be defined by the action as in a CF for two people walking and talking, where it would be opened by a greeting protocol and closed by a goodbye protocol and, without any marked boundary, would simply include the two people.

Inhabitants of a coaction field are either real inhabitants or virtual inhabitants. A real inhabitant (RI) is a real human, animal, insect or plant. A virtual inhabitant (VI) consists of the following:

- *ICA (Inhabitant shot by camera)*: Real people, animals, insects or plants shot with cameras, (transmitted) and displayed in VR.
- *ICV (Inhabitant recognised by computer vision)*: Real people, animals, insects or plants recognised by computer vision, (transmitted), reproduced by computer graphics and displayed using VR.
- *ICG (Inhabitant generated by computer graphics)*: An imaginary or generic

9

life form created by computer graphics, which may have human or artificial intelligence, (transmitted) and displayed in VR.

A VI is described as: (ICA, ICG, ICV). The term avatar is now popularly used for ICG and ICV.

Again we can see that ICA and ICV are derived from referents in the real world whereas an ICG is imaginary or generic. By generic is meant some standardised, abstracted non-specific version of a concept such as a man or a woman or a tree. It is possible to modify VR derived from RW or mix it with VR derived from SCG. For example, it would be possible to take a person's avatar that has been derived from their real appearance and make it slimmer, better looking and with hair that changes colour according to mood. Making an avatar that is a good likeness can take time. A quick way is to take a standard body and, as it were, paste onto it a picture of a person's face derived from a photo.

An ICG is an agent that is capable of acting intelligently and of communicating and solving problems. The intelligence can be that of a human or it can be an artificial intelligence based on neural network, knowledge base, language understanding, computer vision and image processing technologies. The implications are that a coaction field is where human and artificial life communicate and interact in pursuit of a joint task.

The means of communication relate to the way that coaction fields in the first place would have reflected light from the real world and projected light from the virtual world. This would permit communication by written words, gestures and by such visual codes as body orientation and actions. They would also have sound derived directly from the real world and from a speaker linked to a computer source that would allow communication by speech, music and coded sounds. In time it will be possible to include haptic and olfactory codes.

The knowledge domain relates to the fact that a coaction field is a purposive system. Its elements function in concert to achieve goals. To do this there must be a shared domain of knowledge. In a CF this resides within the computer-based system as well as within the participating inhabitants. A conventional game of tennis is a system whose boundaries are defined by the tennis court. The other elements of the system, such as balls and rackets, become purposively interactive only when there are players who know the object of the game and how to play it. Intelligence resides in the players. However, in a virtual game of tennis all the elements, including the court, the balls, the rackets and the Net, reside in a database. So too do the rules of tennis. A CF for HyperTennis combines the two. The players must know the game of tennis and so too must the computer-based version of the system. It would be difficult to cheat.

This brings us to the laws in a coaction field. These follow the laws of humans and the laws of nature. By the laws of nature are meant the known laws of physics, biology and chemistry. These are, of course, a given in that part of a CF that pertains to the real world. They can also be applied to the intersecting virtual world, but this does not necessarily have to be the case. For example, moving objects may behave as they would in physical reality and change shape when they collide. Plants can grow and bloom and seed and react to sunlight naturally. On the other hand, things can fall upwards in VR and plants can be programmed to grow in response to music. These latter are examples of laws devised by humans that could be applied to the virtual aspect of a CF. However, human laws relating to socially permissible behaviour will also apply in the real-world part of a CF. Some readjustment of these laws, as is happening with the Internet, may become necessary as HyperReality expands as an infrastructure technology.

Besides the social laws, there are also the behaviour protocols that apply to a particular CF and, as in the real world, these will need to be defined. How does one enter into and behave in a CF for tennis, translation, nursing and education? Is it the same as in a conventional system for these activities?

Coaction fields are systems and to work they must have control functions. These include:

- Common area controls for sharing knowledge.
- A support control for interpretation and translation between different languages as well as interpretation of words, gestures and diagrammatic representations.
- Controls for applying the laws of biology, physics and any special laws.
- Controls for the creation, integration and separation of coaction fields according to the following rules:
  - A coaction field is created before any coaction takes place.
  - A coaction field can be defined within another coaction field.
  - A coaction field within another coaction field can have newly defined communication means and rules. Attributes defined by the outer coaction field may be inherited if they do not conflict with those defined for the inner coaction field, e.g. CFx and CFy may merge into CFxy and subsequently become separated and restored to their original condition as coaction fields. The attributes of CFxy are then determined according to the following rules:
    - When there are conflicting attributes, only one is chosen.
    - When there is no conflict, both attributes coexist.

In Figure 1.3 two adults, one real and one virtual are discussing something in CFa, which is a coaction field for interpreting between Japanese and English. They must be able to speak either Japanese or English. A real boy is playing ball with a virtual puppy in CFc. The boy and the puppy share the knowledge of how to play ball. A virtual girl is showing her virtual balloon to a real girl in CFb.

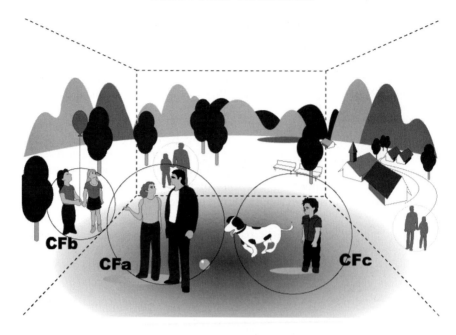

Figure 1.3 Example of HR.

If we assume that the ball in CFc rolls into CFb and that the two girls begin to play ball with the boy and the puppy, then fields CFc and CFb become integrated to form a new coaction field CFbc. However, if the boy does not like playing with the girls and will only play with the puppy, then CFb and CFc become separated again. All of them are in a HyperWorld.

RELATED TECHNOLOGIES

The term 'hyper' in HyperReality is used because HR is more than the sum of physical reality and virtual reality. It is predicated on systematic interaction between the two component realities. Because of this, it is a new form of reality that, as we shall see in this book, has attributes above and beyond its component realities. Of course HR is only possible because of the development of virtual reality, in particular the development of Distributed Virtual Reality (DVR). However, the main thrust in VR has been to generate realities in their own right, which are to be entered into on their own terms and which are separated from the real world. Exceptions to this are Augmented Reality (AR) and Mixed Reality (MR).

Distributed Virtual Reality

Initially, computer-generated virtual realities were experienced by individuals at single sites. Then sites were linked together so that several people could interact in the same virtual reality. The development of the Internet and broadband communications now allows people in different locations to come together in a computer-generated virtual space and to interact to carry out some co-operative purpose. This is Distributed Virtual Reality (DVR). HyperReality incorporates DVR, but it also links DVR with the real world in a way that seeks to be as seamless as possible.

Augmented Reality

Augmented Reality (AR) is fundamentally about augmenting human perception by making it possible to sense information not normally detected by the human sensory system (Azuma 1997). A 3D virtual reality derived from cameras reading infrared or ultrasound images would be AR. A 3D image of a real person based on conventional camera imaging that also shows images of their liver or kidneys derived from an ultrasound scan is also a form of AR. In that this last example involves the insertion of a VR that humans would not normally be able to see, into a VR of a physical setting as a human would see it, it could be confused with HR. They are, however, different. In AR the key concern in relating a VR derived from a conventional camera view of physical reality with a VR based on an ultrascan of physical reality is one of registration – the liver or kidney has to correctly fit the person's body and the relationship between the person's body and the superimposed VR of the organ does not change when the person moves. In HR, it is the real and virtual elements that interact and, in doing so, they change their position relative to each other. Moreover, the interaction of the real and virtual elements can involve intelligent behaviour between the two and this can include the interaction of human and artificial intelligence.

Mixed Reality

The term Mixed Reality (MR) has received some favour as a coverall term for having virtual objects and real objects available within the same visual display environments (Milgram and Kishino 1994). The term is taken to include a very broad continuum ranging from videoconferencing to augmented reality and the mixture of different concepts of physical and virtual realities (ibid.). It does not, however, seem to refer to the interaction of physical and virtual reality, nor are mixed and augmented reality seen as places for the interaction of human and artificial intelligence.

APPLICATIONS OF HYPERREALITY

The applications of HR would seem to involve almost every aspect of human life, justifying the idea of HR becoming an infrastructure technology. They range from providing home care and medical treatment for the elderly in ageing societies, to automobile design, global education, games and recreational activities. The second half of this book looks at some of these. Below are some examples of applications.

Telemedicine

HyperReality provides a means for telemedicine – the delivery of medical services at a distance. Nurses and doctors can monitor the condition of patients without having to physically meet them. The system could be used to televisit housebound patients or for consultancies between GPs and specialists.

Figure 1.4 shows a HyperClinic. The doctor and the patient are in different places. The doctor can visually check the patient and talk with them, and a certain amount of telemetering of such things as blood pressure, temperature and pulse is possible.

Figure 1.4 A HyperClinic.

Art galleries and the HyperMuseum

Figure 1.5 suggests how an art gallery or a museum might function in HyperReality. It would involve creating a database (DB) of virtual replicas

14

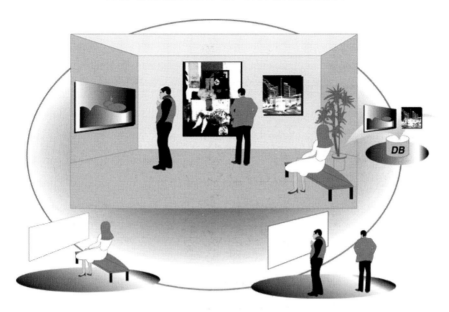

Figure 1.5 A HyperGallery.

of the items in a collection. This is currently being done with a collection of cultural artefacts at the Tokyo University of Fine Arts. In Figure 1.5 a couple of friends and a bedridden woman are visiting the gallery. In reality they are at locations remote from the gallery. Their human forms were recognised beforehand by computer vision technologies. They can move through the gallery by manipulating their avatars using either hand gestures or voice control or a keyboard.

Manufacturing on demand

Figure 1.6 illustrates how manufacturing on demand can be made possible through HR. Besides being used in the automotive industry to design and manufacture personalised cars, it could be used in a similar manner to design and manufacture such things as personalised refrigerators, sound consoles or video systems.

In Figure 1.6 a potential car buyer is linked in a coaction field with two designer/salespeople to discuss the kind of new car the buyer would like to have. The database (DB) holds basic car models and components and the knowledge necessary for designing a car. The designers are showing a proto-type design to the customer and can change the colour and the shape of the car according to the customer's wishes before the car is manufactured.

Such an interaction does not necessarily require human beings on the design side. The designers could be agents generated by a computer and

Figure 1.6 Designing a personalised car with HR.

programmed to interact with a customer over their design needs. This would take place in a coaction field where humans and agents share a common knowledge of cars and their features. To help him conceptualise what he wants, the customer can handle the 3D model of a car and see it from every aspect. He can open doors, sit in the driver's seat and take the car for a drive. Finally, after checking that the car has all the factors the customer wants, the design information can be sent to the factory and the car manufactured.

Designing and furnishing a house

In much the same way, it will be possible with HR to buy and furnish a house by examining VR models of houses with estate agents and architects. It would be possible, for example, to walk through the rooms of a real house with designers, and call up VRs of alternative possibilities for doors, windows and walls. Different colour schemes, curtains, furnishings and furniture could also be tried out.

Acting

People can be audience or actors. They can enter on to a stage that may be real or virtual and interact with real or virtual actors or actors who are intelligent agents acting according to a shared knowledge of the play (Tiffin and Rajasingham 1995). It would be possible to stage anything from Matsu-no-Roka corridor theatre to Shakespeare in the Globe Theatre.

Travelling

Using HR technology, people can go anywhere they want without leaving the place where they live. Someone in New Zealand could visit Jou Jakkou-ji temple in Japan and walk through it without actually having to leave their own country. To make this possible, shots of Jou Jakkou-ji temple would be taken by camera and the shapes and colours of the temple recognised by computer vision and stored in a computer. The information would then be transmitted to New Zealand. The scenes would then be reconstructed and displayed on the screen stereoscopically through VR technology so that the New Zealand viewer could enter and walk through the temple.

Education

The first field of application for HyperReality is that of education and a coaction field called a HyperClass is being developed. A HyperClass is a 3D space where virtual and real teachers and students are brought together through communication links, to hold classes and work together as if they were gathered in the same place. Figures 1.7 and 1.8 illustrate the concept.

A prototype HyperClass system has been developed and experiments have

Figure 1.7 Image of a HyperClass. Real people, virtual people and virtual objects interact in a coaction field.

17

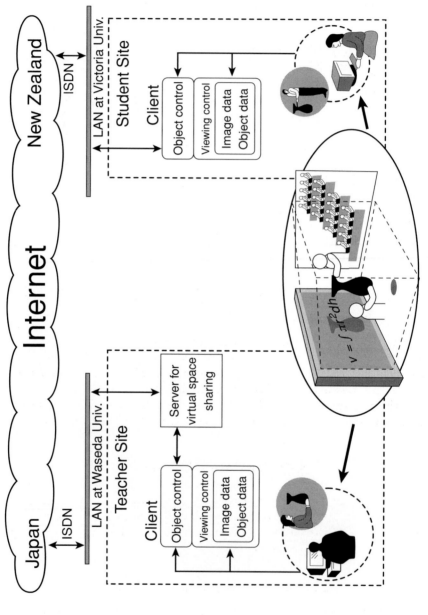

Figure 1.8 Basic relationships between HyperClass sites in Japan and New Zealand.

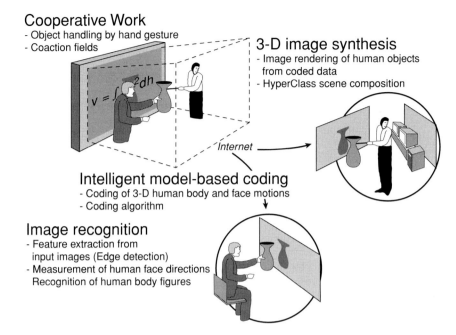

Cooperative Work
- Object handling by hand gesture
- Coaction fields

$$v = \int^2 dh$$

3-D image synthesis
- Image rendering of human objects from coded data
- HyperClass scene composition

Internet

Intelligent model-based coding
- Coding of 3-D human body and face motions
- Coding algorithm

Image recognition
- Feature extraction from input images (Edge detection)
- Measurement of human face directions Recognition of human body figures

Figure 1.9 Systems configuration of the HyperClass experiment. The student site in New Zealand is linked through the Local Area Network of Victoria University via the internet to the teacher site at Waseda University.

been carried out interconnecting Waseda University in Japan with Victoria University of Wellington in New Zealand via the Internet (Terashima and Tiffin 1999; Terashima, Tiffin and Rajasingham 1999; Terashima 1998). Figure 1.9 gives details of the systems configuration used.

In the experiments, people in both countries handled virtual objects such as those in Figure 1.10 that illustrate Japanese cultural heritage.

KEY TECHNOLOGIES

The principles of HyperReality have been demonstrated in the laboratory, but for HR to move into common practice and become an infrastructure technology, a number of key technologies need to be developed, especially in the field of computer vision.

Computer recognition of objects

One of the key technologies for HR development is computer recognition of objects. The problem is that many real objects have some kind of movement. A human may be walking, running or playing, while leaves on a tree behind

Figure 1.10 Samples of Japanese cultural heritage used in the HyperClass experiment. These are ancient ceramic objects, part of a collection at Tokyo University of Fine Arts, that are being copied as virtual objects to be stored in a database.

them may be moving in the wind. Recognising complex moving objects presents problems, especially when inhabitants are coacting in a coaction field where some knowledge is shared between them and where the laws of physics and biology apply.

Today, in order to display images of a HyperWorld from the observer's perspective, cameras are placed around the targeted natural and physical objects, and a method for switching positions of the cameras according to the observer's perspective is being considered. However, this method does not offer sufficient realism because continuity is lost when cameras are switched. To overcome these problems, the images of targeted natural and physical objects are first placed into the computer using Computer Vision. The images can then be reproduced from the observer's perspective and displayed in real time. To enable this, technologies are required that can detect the observer's perspective and recognise, reproduce and display images from that perspective.

Figure 1.11 illustrates the above technologies as they are used in relation to human figures. A human figure is first recognised and made into a wireframe model. Then a texture is obtained and stored in the workstation. Human figures are easy to model because they have many features in common, but such objects as trees are difficult to model because they can vary greatly in shape. A generic model of a person can be created and used like a mannequin as a basis on which to 'dress' the appearance of an individual. It then becomes an easily recognisable target for computer recognition. However, new technologies are needed to recognise physical objects that are difficult to model.

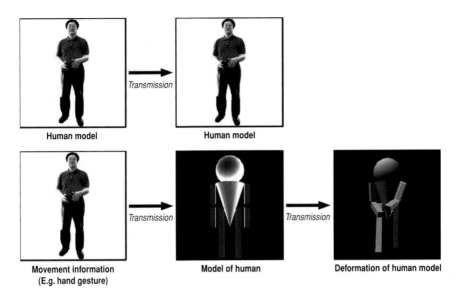

Figure 1.11 (Top) conventional image transmission by videoconferencing, and (bottom) image transmission by coding technology.

Recognising movement information

In order to synthesise a human image from a real human, information also needs to be detected in real time about head, hand and other body movements. Research is currently under way on non-contact movement detection methods as well as on detection methods in which sensors are placed on numerous parts of the human body. Movement information from other natural physical objects may not be so easy to detect. To target a single tree, for example, the movement of each individual branch, twig and leaf needs to be detected.

Touch and heft

In the real world, we feel the texture of an object when we touch it, and when we grasp it, we feel its weight, referred to as heft. To give a sense of realism, these feelings have to be accomplished in VR. There is research on touch and heft. For example, a system has been developed, using a torque motor, that allows viewers to feel the heft of the object when they grasp it. However, for this to happen they have to use the arm of the system to touch the object. More convenient methods need to be developed in the future.

21

Creating a sense of realism

Even if virtual reality can be inserted into physical reality without some kind of bezel separating them, with today's technology it is obvious what is real and what is virtual because of the difference in quality of the images and movements of the virtual components. Accompanying sound lacks directionality, and touch and heft in virtual reality are unconvincing. People are sensitive to very small differences in lip sync in a film. How much greater then is the problem of synchronising different kinds of visual, auditory and haptic information? High-quality synthesising technology is needed to integrate the different sensory aspects of HR.

Displaying images

When images are displayed on a screen, a three-dimensional effect can be obtained by viewing the image with shutter glasses, where images corresponding to the left and right lenses are switched and displayed at high speed. To see stereoscopically a lenticular screen (Figure 1.12) displays images, corresponding to the left and right eye viewpoints, through 3.6 mm semicylindrical plastic plates. However, as yet the results do not give

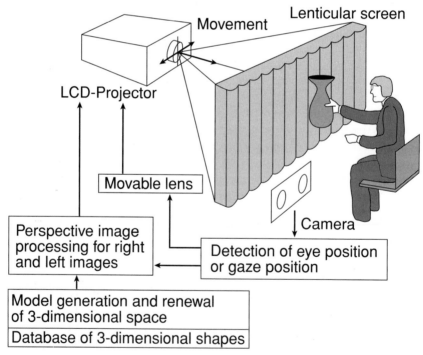

Figure 1.12 Stereoscopic display system employing eye-position tracking.

a natural-seeming stereoscopic 3D image and any rapid head movement by the observer results in the loss of three-dimensionality. At the moment such systems are partly mechanical and a wholly electronic system may resolve some of the problems.

Protocol

Protocols have to be established for people entering and communicating in HyperReality together. HR protocol has two syntaxes: an abstract level and a concrete level.

At the abstract level, HR protocol is the same as Open Systems Interconnection (OSI) protocol, which has seven layers: physical/electrical, data link, network, transport, session, presentation and application level protocols. OSI is used for the interconnection of terminals and/or computers in consistent ways.

HR protocol is used for the interaction of inhabitants in a coaction field. At the transport level, a joint operation is performed where inhabitants, real or unreal, join together and a coaction field is created. Next, they establish the session through the session level protocol. Following this, they are able to play or work together.

In the concrete syntax, HR protocol is performed as follows: (1) a user selects from a menu of coaction fields for specific domains of interaction; (2) when two or more people make the same selection, a coaction field is created; (3) they then negotiate how they will carry out the joint interaction and a coaction field is established for the duration of the interaction; (4) when the interaction is completed, the coaction field is deleted.

CONCLUSION

Virtual reality is in its infancy. It is comparable to the state of radio transmission in the last year of the nineteenth century. It worked, but what exactly was it and how could it be used? The British saw radio as a means of contacting their navy by morse code and so of holding their empire together. No one in 1899 foresaw its use, first for the transmission of voices and music, and then for television. Soon radio will be used for transmitting virtual reality and one of the modes of HR in the future will be based on broadband radio transmissions.

This chapter has tried to say what HR is in terms of how it functions and how it relates to other branches of VR. It has also briefly alluded to its history. HR is still in the hands of the technicians and it is still in the laboratory for improvement after trials. But a new phase has just begun. HyperReality is a medium and the artists have been invited in to see what they can make of it. Making meaning out of HyperReality is the subject of the next chapter.

REFERENCES

Azuma, R.T. (1997) 'A survey of Augmented Reality', *Presence, Teleoperators and Virtual Environments*, 6, 4: 355–385.

Bajura, M., Fuchs, H. and Ohbuchi, R. (1992) 'Merging virtual objects with the real world: seeing ultrasound imagery within the patient', *Computer Graphics*, 26, 2: 203–210.

Milgram, P. and Kishino, F. (1994) 'A taxonomy of Mixed Reality visual displays', *IEICE Transactions on Information Systems*, E77-D, 12: 1321–1329.

Terashima, N. (1993) 'Telesensation: a new concept for future telecommunications', *Proc. TAO First International Conference on 3D Image and Communication Technologies.*

Terashima, N. (1994a) 'Virtual space teleconferencing system', *Proc. 3rd International Conference on Broadband Islands.*

Terashima, N. (1994b) 'Virtual space teleconferencing system: a distributed virtual environment', *Proc. IFIP World Congress '94.*

Terashima, N. (1995a) 'HyperReality', *Proc. International Conference on Recent Advance in Mechatronics.*

Terashima, N. (1995b) 'Telesensation: fusion of multimedia information and information highways', *Proc. International Conference on Information Systems and Management of Data.*

Terashima, N. (1998) 'HyperClass: an advanced distance education platform', *Proc. IFIP Teleteaching '98.*

Terashima, N. and Tiffin, J. (1999) 'An experiment of virtual space distance learning systems', *Proc. Annual Conference of Pacific Telecommunication Council.*

Terashima, N., Ohya, J., Kitamura, Y., Takemura, H., Ishii, H. and Kishino, F. (1995) 'Virtual space teleconferencing: real time representation of 3D human images', *Journal of Virtual Communication and Image Representation*, 6, 1: 1–25.

Terashima, N., Tiffin, J. and Rajasingham, L. (1999) 'An experimental distance education system using objects of cultural heritage', *Proc. IEEE Multimedia Systems Conference.*

Tiffin, J. and Rajasingham, L. (1995) *In Search of the Virtual Class: Education in an Information Society*, London: Routledge.

2

THE HYPERREALITY PARADIGM

John Tiffin

Editors' introduction

The late Rudy Bretz told a story of how, as an inventive cinematographer in the mid-forties, he was invited by one of the big American broadcasting corporations to help them figure out some new equipment that had just arrived. It was called television. Everything was in crates, there was little by way of precedent as to how to organise it or use it or, indeed, what to use it for. So Rudy worked out a system for making television programmes and later wrote it all up in a book, which became a bible for producers (Bretz 1953). Like the Bible *it creates a comprehensive explanation of its subject. Such a disciplinary matrix of principles, problems, methods, concepts and standards of evaluation applied to the field of science was defined as a paradigm by Thomas Kuhn in a postcript to the 1969 edition of his* The Structure of Scientific Revolutions. *Today HyperReality is at the stage television was at when John Logie Baird was experimenting with it: a technological capability waiting to be developed but still in the laboratory. It is a medium for communication between the real and the virtual, between human and artificial intelligence and between fact and fiction. But it is a medium without messages. It has not yet been used in anger. It has done no harm and it has done no good and the bible on how it should be used has yet to be written.*

To make a message in a medium requires a code, which then exists in two dimensions: the abstract plane of theoretical possibilities it offers and the real life plane of its actual usage. Saussure (1916) recognised the difference when he spoke of 'langue', by which he meant language in the abstract with all that it is capable of, and 'parole' meaning language as it is actually used. The terms 'paradigm' and 'syntagm' are used in communications studies in a similar way to langue and parole for the dimensions of possibility and practice that intersect to make meaning in any medium.

The first and second use of the term paradigm in this introduction are not discordant. The paradigmatic dimension of a medium contains the principles, methods, concepts, rules and evaluative standards by which syntagmatic messages are made. This chapter is concerned with the paradigmatic dimension of HyperReality as a medium and the possibilities it presents in the abstract for making meaning. In

25

doing this it seeks to discriminate HR from competing and complementing media. This raises the Khunian issue implicit in the subtitle of this book: could HyperReality become the dominant medium by which we know reality?

COMPETING AND COMPLEMENTING PARADIGMS

paradigm: the example or pattern on which something is based

(OED)

Like the choice between competing political institutions, that between competing paradigms proves to be a choice between incompatible modes of community life. Because it has that character, the choice is not and cannot be determined merely by the evaluative procedures characteristic of normal science, for these depend in part upon a particular paradigm, and that paradigm is at issue. When paradigms enter, as they must, into a debate about paradigm choice, their role is necessarily circular. Each group uses its own paradigm to argue in that paradigm's defence.

(Kuhn 1962)

According to Terashima (Chapter 1 this volume) HR is the seamless inter-lacing of virtual reality with physical reality and of artificial intelligence with human intelligence, an interlacing that is becoming possible with advanced telecommunications and computer technologies. From my perspective as a communications specialist this means that virtual people, virtual objects and virtual settings can interact and thereby communicate with real people, real objects and real settings as though they were all part of the same world. What is made real by the dance of atoms can coexist with what seems real from the sensory play of bits of information on the receptor nerves of human bodies. As time goes by, the difference between what is really real, and what exists in effect, but not in fact, could blur. People could come to live in a world in which they cannot readily distinguish whether what they see, hear, smell and touch is derived from the physical world or mediated by information technology. It can be argued that this is no more than an acceleration of a trend that can be traced back to story-telling by the light of a campfire in a cave with some visual aids on the wall; that thousands of years ago people were already using machine intelligence for such things as telling the time, the date and the level of water in a river. What will be new in HyperReality is the autonomy of the real and virtual elements and the human and artificial intelligences in their communication with each other. It will be as though the painted things came off the cave wall to join the group around the fire and argue with the storyteller.

As with mobile phones, the advantages of HR seem immediate and obvious and the disadvantages may only be recognised in retrospect. There is hardly any communicative aspect of our lives that, at first thought, could not be improved by HR. Education, medicine, business, government, shopping, entertainment and the whole gamut of social activities that constitute a society could all be conducted in HyperReality. One day it is possible that they will be.

Jayaweera (1987) sees the early inventive stage of a technology as neutral. It is at a later stage, when it becomes organised and exploited within the economic, social and political structures of society, that it becomes an instrument of power. Clearly HyperReality is in the first inventive stage of this model. It is still malleable. The enabling infrastructure of broadband telecommunications and powerful computers is not in place. It has the potential to become a pervasive unifying communication infrastructure that could define the modus vivendi of the information society, but for that to occur it will need to compete and adapt to other contending communications media.

What do we mean by the information society? Daniel Bell (1976) used the example of the US to classify the development of society into three phases: the pre-industrial, the industrial and the post-industrial. He argued that, when the majority of workers in America were involved in manufacturing, it was, ergo, an industrial society. Prior to that, around 1860, when most workers were employed on the land in farming, forestry and mining, America could be regarded as pre-industrial. When, as was the case from the 1960s, the main field of employment became the service industries, then America became a post-industrial society. There have been many attempts to define and explain what comes after the industrial society. Toffler (1970) called it the 'third wave' (the second wave equates with the industrial society and the first wave with the pre-industrial society). Beniger (1986) thought of it in terms of control. Castells (1997) refers to it as the 'network society'. Of special relevance is the work by Dordick and Wang (1993) who see the information society in terms of infrastructure. An information society cannot function without access to information. Such an approach aligns with that of Bell. The majority of workers in the US could not have worked in factories if it had not been for the railway madness that, in the decade preceding 1860, created an enabling infrastructure across the US that transported raw materials to factories and finished goods to markets. Something similar appears to be taking place now with the Internet. It is putting in place an infrastructure for the transport of information that will allow anyone, anywhere, to work, play, buy, sell, study, write and do just about anything that in some way involves processing information and not atoms. The more people use it, the more it takes over their lives and their way of communicating. It is this pervasiveness that Kranzberg and Pursell (1967) see as the hallmark of technological revolutions.

If ever there was a defining moment for becoming part of an industrial society it was when the railway came to town. It is tempting to see the day a household jacks into the Internet as the point at which it joins the information society. Yet, unlike the railways, which have changed in concept very little, the Internet is not a stable technology or even a stable concept, nor is it clear when it will become stable or what shape it will have when it does. The so-called superhighways of megabandwidth, with the promised potential to radically change the Internet, are not yet in place. The Internet has barely begun to use radio or to link into mobile phone and Low Earth Orbiting (LEO) satellite technologies. The principal inter-face device is still the QWERTY keyboard, designed with maximum inefficiency over a hundred years ago so that the striking keys on a type-writer would not jam.

The steam engine and later the internal combustion engine may have shaped the physical form of the industrial society, but it was mass media that made the mental landscape of the industrial society. Still today mass media provide the people of the world with their paradigm of reality. The rapid rise of the Internet in the last decade of the twentieth century can be seen as challenging this. HR is currently being adapted for use on the Internet and it may be that its future will be bound up with that of the Internet as it becomes broadband and radio-supplemented and accessible by voice, touch and gesture. In this case HR can be seen as complementary to the Internet but competing with mass media.

HyperReality arises from a cluster of innovative technologies. In turn HR is part of a wider cluster of emergent technologies. However, in the sum of its possibilities HR as it develops will take on a life of its own and become what Winner (1977) calls an autonomous technology. It could then find itself in competition with the Internet. Alternatively the Internet may subsume HR or vice versa. Castells gives some indication of how complex the move from theory to praxis in a technology can be:

> Technological breakthroughs came in clusters, interacting with each other in a process of increasing returns. Whichever condition determined such clustering, the key lesson to be retained is that technological inno-vation is not an isolated instance. It reflects a given state of knowledge, a particular institutional and industrial environment, a certain availability of skills to define a technical problem and solve it, an economic mentality to make such application cost-efficient and a network of producers and users who can communicate their experiences cumulatively, learning by using and by doing: elites learn by doing thereby modifying the appli-cations of the technology, while most people learn by using, thus remaining within the constraints of the packaging of the technology. The inter-activity of systems of technological innovation and their dependence on certain milieux of exchange of ideas, problems and solutions are critical

features that can be generalised from the experience of past revolutions to the present.

(1997: 37)

Certainly this seems true of mass media. The application of steam power to printing interacted with the growth of railways and macadamised roads and the development of universal education to magnify the one-to-many characteristic of print. Similarly, the growth in world telecommunications has extended the one-to-many aspect of radio and television to a point at which a handful of production centres provide programmes for most of the world. Throughout much of the twentieth century critical theory, with its roots in Marxism, cultural studies and feminism, has sought to draw attention to the consequent development of cultural hegemony. The Internet, in contrast, is a fully meshed many-to-many medium that obscures the difference between doer and user. Unlike mass media, Internet message production is not something controlled by a small elite with the resources for the large capital investment involved. The Internet may create new elites, but it does not by its nature privilege existing elites.

An International Congress on Education Technology and Change held in Cali in Columbia in 1999 was organised by a group of people from Choco, the poorest state in the country, around the theme of 'The Virtual University'. They saw hope for their state in this idea. A virtual university could lead the way in researching how they could use the Internet and in teaching teachers who would teach people to make a living from the Internet. Colombia has become a pariah country because of its reputation for drugs and violence. Nobody wants to go there or have ships arrive with its produce. The collapse of exports and the tourist industry has driven the country into recession and ordinary people into poverty. As an aspiring industrial society, Colombia is going backwards – nowhere more so than in Choco. They see the Internet as an answer to economic isolation and a chance to become an information society.

In Latin America there is a saying that technology is a knife with two edges that can cut both ways. At the beginning of the twentieth century Kuwait and Saudi Arabia were among the poorest countries. Because of the development of motorised transport in other countries through the century, they are now among the richest. Half a century ago, when 'coal was king' and powered the factories and railways, the industrial heartlands of Europe, Japan and North America were settings for a welfarism in which poverty was almost eliminated. Now these are industrial wastelands and poverty is again rife. We cannot as yet imagine the winners and losers with the Internet and HyperReality, nor what random butterfly effect may occur as HR develops.

HyperReality lies somewhere in between the extremes of mass media and the Internet. Like the Internet it is a fully meshed many-to-many medium.

It will be possible for anyone anywhere to become a participant by entering a coaction field and, unlike the mass media context, they can interact with other components and change what happens. However, in the creation of coaction field databases we find an example of what Castells, quoted above, calls the 'constraints of the packaging of the technology'. Coaction field production is likely to be every bit as specialised, complex and expensive as television and film production. After all, films and television programmes, like radio programmes, books and articles in newspapers and magazines, are similar to coaction fields in that they are paradigmatic. As in a coaction field for HyperFootball, a newspaper report or television programme on a game of football assumes that the reader/viewer understands the rules and concepts of football and the language used. Although they cannot physically participate in the game, the reader or viewer can, in their mind, create a virtual reality of the game in which they can take part and even change the result.

A key difference between the Internet and mass media lies in the contrast between the linearity of mass media, where a message has a beginning, a middle and an end in that order, and the hypertext capability of the Internet, where it is possible to surf from one text to another at the whim of the user. The nature of HR aligns it more with mass media. It will be possible to surf, in the sense of flipping from one coaction field to another, but this is more like clicking through television channels or flipping through the pages of a newspaper. The hypertext capability of the Internet allows for surfing with sense. A user can make links that are logical and search for patterns of meaning in networks of web sites. This is possible because they can progress at their own pace in a medium that is still essentially text- and picture-based and therefore asynchronous. HR is, by contrast, synchronous in nature. The components of a coaction field come together to interact purposively in real time. Some coaction fields such as those for games will have a determined duration and begin at a fixed time. Such coaction sites would be fixed and focused events and access may need to be restricted and paid for. The surfing hypertext mode does not fit with such activities. The Internet came into existence as a postal and library service that was more efficient and less expensive than the ones that already existed, and was easily available to anyone with a PC and a telephone. However, HR is a medium for happenings and meetings. Although access to HR on the Internet may cost little more than that to the Internet itself, access to specific coaction fields could be in the order of access to real-life events. A sports broadcast or a press report on a game are only possible because a game actually takes place and the same is true in HR.

HR also needs to be distinguished from VR. HyperReality includes virtual reality but is different from VR per se. In the popular image, virtual reality is a technology that provides computer-generated realities that are an alternative to physical reality. Head-mounted display units (HMDUs) exclude images that are derived from the real world. Instead they provide images that are generated from information in the database of a computer. The person wearing the HMDU gets a sequence of images from a computer, consistent with being inside a three-dimensional world. The changing 3D image can also be matched by changing 3D sound. Just as images in an HMDU address each eye, to give perspective, so sound in an HMDU addresses each ear. As the head is turned, sound and image change accordingly. The addition of data gloves makes it possible to see and use hands in virtual reality. Force-feedback provides a sense of touch that corresponds to what the virtual hands do. The development of data-suits will provide computer-generated tactile sensation to the whole body, corresponding to what is seen and heard in virtual reality. This is the technological trend towards what Marvin Krueger, in a conference in Gothenberg in 1994, called 'body-condom virtual reality'. It seeks to replace stimuli derived from the physical world with stimuli generated by a computer.

HyperReality is a development in a different direction. It seeks to make virtual reality something that is experienced as part of physical reality, so that virtual and real phenomena appear to interact with each other: HR is VR *as well as, not instead of,* PR. The developmental thrust in total immersion VR is to provide virtual worlds that are more and more convincing to those who step inside them. In HR, however, the research trend is towards providing HyperWorlds where the commingling of what is virtual and what is real is seamless and so appears natural.

The difference between VR and HR is like the difference between the cinema and the telephone. The cinema can be thought of as an early attempt at full immersion VR: when the lights go down the audience stop speaking to each other and become oblivious to the surrounding physical reality as they allow the VR projected in front of them to take over their perceptual system. In contrast, the telephone could be seen as a precursor to HyperReality. The two people telephoning conceive of themselves as real and the other person as virtual. Who is real and who is virtual depends on where each of them is. Yet, as a telephone conversation proceeds, the interaction seems natural. There is a sense of shared aural presence in the real and virtual voices. A simple way of thinking about HR is that we will be 'telephoning' our whole bodies and that we will be able to see, hear, touch and even smell the people we are 'telephoning'. In addition, with HR you can also telephone your context: your cat, your piano or the view from your window.

The terms distal and proximal refer to the beginning and end of the stimuli that bombard our bodies. Distal stimuli are the smoke as it comes from a burning frying pan, the scream as it articulates in the mouth of a child, the picture as it exists on the screen of a television set. Stimuli are proximal when they impact on the receptor nerves of the human sensory equipment: the smell of burning in the nose, the agitated sound waves of a scream as they whirl into the ear, photons from a television screen at the point where they impact on the fovea of an eye. What we are conscious of, what we strive to make meaning from, are the proximal stimuli that impinge on our neural system. Whether the distal origins of stimuli are in physical reality or virtual reality is inferred by matching our internal accumulated knowledge of the world to incoming proximal stimuli. How many of us have answered a ringing telephone only to discover that the sound came from a television or radio programme? We could not distinguish whether the distal stimulus was from the physical reality of a real phone or the virtual phone simulated in a broadcast. HyperReality is when you can answer the phone whether it is real or virtual.

We can easily distinguish a telephone voice from a real voice because the limited bandwidth in today's telephony means that the normal range of a voice is flattened. In the same way, it will at first be easy to distinguish the virtual from the real in HyperReality. Virtual things will lack definition and detail, movement will be jerky and unnatural and in coaction with what is real, virtual sound, movement and touch will seem 'out of sync'. We will probably come to accept the virtual reality components as we accept the voice on the telephone, but the challenge to the technology as it develops will be to achieve sensory verisimilitude. It will in time become increasingly difficult to distinguish the virtual from the real. Could there come a day when the only way to tell whether something is made of atoms or bits of information will be to try to eat it? Will we pass that point and find it becoming easier to recognise the virtual because it is clearer, more intense, more interesting and, in effect, better than physical reality?

We already live in a mixture of the real and the virtual. We daydream while driving on roads that are full of cars driven by other daydreamers. However, the virtual realities generated outside ourselves are normally separated from our physical surroundings by some kind of frame. The text of a book is framed by its pages. Television is contained in a three by four frame. There are frames around the pictures on a wall. The cinema screen frames a virtual world. The first widespread use of HR will be on the Internet where it will be framed within the monitor of a computer. The long-term goal of HR research, however, is that the frames will disappear and we will cease to be conscious of any seams between the virtual and the real. It is possible that one day the person we dance with, or the Godzilla doll we let the children play with, may be virtual and this no more concerns us than it would today if our children were playing with a toy and our partner reading a romantic book about Rudolph Valentino.

Rooms in modern urban societies create an artificial reality in which we can safely enter virtual realities. Books on a shelf or magazines on a table are virtual reality hot spots in the physical reality of a room. Pick one up, open it and start reading and you drift into a virtual reality that can be so enveloping that you no longer notice your surroundings. On the other hand, you can flip through a newspaper while carrying on a conversation. You can switch off the lights and sit in front of a large-screen television set and become deeply absorbed in it or the television can murmur away to itself, little more than a window of moving wallpaper, while you go about your business with half an ear cocked for any salience in its wasteland. Essentially, however, you are aware that the television is television, the book is a book, the Internet is the Internet and the virtual realities they invoke are distinct and separate from the reality of the room. When the room becomes a space for HyperReality this will change.

In suggesting that the telephone can be thought of as a primitive form of HR, it needs to be remembered that telephony developed as an analog technology. A telephoned voice corresponds to and is directly derived from a real voice. This, along with the speed at which telephone calls are transmitted, allows direct synchronous communication between people. HyperReality, however, is a digital technology. Communication in HR is mediated by computers. The communication process may seem in its apparent synchronicity to be a direct transaction like telephoning, but between the sender and the destination of a message in HR there lies a computer. That computer can be programmed to change the message in many, many ways and to participate in the communication. This brings a dimension to communication between people that has not existed before in any medium. A HyperWorld is not only where what is real and what is virtual interact, it is where human intelligence meets artificial intelligence. It is this that warrants the prefix 'Hyper' used in its sense of 'more than' (*OED*).

THE COACTION FIELD AS A PARADIGMATIC 'SHELL'

All messages, therefore, involve selection from a paradigm and combination into a syntagm. All the units in a paradigm must share characteristics that determine membership of the paradigm, thus letters in the alphabetical paradigm, numbers in the numerical paradigm and notes in the musical paradigm. Each unit in the paradigm must be clearly differentiable from other units: it must be characterised by distinctive features. Just as the paradigm is governed by shared characteristics and distinctive features, the syntagm is determined by rules and conventions by which the combination of paradigms is made – rules of grammar and syntax.

(Watson 1993)

The term 'paradigm' in the study of communications refers to a set of semantic elements that can be combined according to a syntax in order to communicate. This is not incompatible with the wider meaning of paradigm given at the beginning of this chapter. A language and a game of cards, in the abstract, constitute paradigms. The purposeful application of elements of a paradigm according to their syntax and semantics constitutes a syntagm. To actually write or speak something or play a game of cards is to create a syntagm. What enables people to engage in some common activity such as talking and writing and playing cards is that they are all agreed on a cards paradigm and a language paradigm and know how to select and use words and cards correctly from them.

All communications can be seen as having paradigmatic and syntagmatic dimensions. When we talk, we are syntagmatically selecting from the paradigm of a language, and the listener is also getting a message from the syntagmatic selection of clothes we choose to wear from the paradigm of dress, and the syntagmatic choice of haircut from the paradigm of hairdressing available to us in our culture. If we are Rastafarians or into Gothic or a member of the Order of St Francis or the Mafia, then our syntagmatic choices of how we dress, behave and speak are made from the Rasta, Gothic, Franciscan and Mafia paradigms. In the educational systems of industrial societies there has been a tendency to paradigmatically compartmentalise communications. We study how to make mathematical syntagms from the mathematics paradigm in a maths class and experiment with syntagmatic expressions of scientific paradigms in laboratories.

Terashima (Chapter 1 this volume) postulates that an attribute of coaction fields is a knowledge domain that is specific to the purpose of the coaction field. The knowledge domain resides in the computer base and the inhabitants of the coaction field and forms a 'dictionary' of the elements of the coaction field and the rules by which they interact. Coaction fields are, therefore, paradigmatic in nature. They provide a potential capability for a domain of interactive communication. They are a technical paradigm 'shell'. In this sense they are similar to banks, churches and schools, which are artificial environments that permit such syntagmatic communications as financial transactions, religious services and lessons from the paradigms of banking, theology and education.

A coaction field, then, represents a paradigmatic set of possibilities for communication that allows a syntagmatic selection for particular purposes. A HyperConference Room allows real and virtual business people to meet. Terashima gives examples of a HyperClinic that allows a patient and doctor to coact in real and virtual dimensions, and a HyperClass that allows virtual students and virtual teachers to coact with real students and real teachers for the purpose of instruction.

Our lives are not so compartmentalised that we speak to our dentists only about our teeth, to our bankers only about financial transactions, to our maths teacher about maths and to our whist partner about whist. In talking to our dentist about the grit in the oyster that broke a tooth, we might discover a joint interest in pearl farming and in asking for a loan from the banker to set up a truffle farm, a common concern with horticulture. On the other hand, a group engaged in a hand of poker could not also operate under the rules of whist. We can syntagmatically draw on paradigms within paradigms provided there are no discordant factors. Similarly, according to Terashima (Chapter 1 this volume) a new coaction field can be created as a result of interaction between existing coaction fields provided that there are no conflicting attributes.

For example, in the case of a doctor examining a patient in a coaction field, there could be a researcher involved in the treatment who would also have a coaction field with the same patient to observe the impact of a medication, and there could be a third coaction field where the researcher and the doctor hold technical discussions concerning the patient. These three dyadic coaction fields could be integrated to form a group coaction field of doctor, researcher and patient linked together at the same time to agree on some changes in treatment. Imagine further that the doctor has a number of students at two different medical schools whom she teaches together in a HyperClass. She arranges for them to monitor in virtual reality both the patient's condition and her discussions with the researcher. We now have a coaction field in which doctor, researcher and students all gather around the patient. Terashima sees no limits to the groups of people/creatures/objects/sets that can come together in coaction fields, provided that the attributes of the integrated coaction fields are not in conflict. There appears to be no conflict between the attributes of the different coaction fields in the example given. They all share the purpose of improving patient health with medical knowledge. However, let us suppose that the patient seeks to enter a coaction field for football. As in a conventional hospital situation, if a patient tried to play football while a consultation was going on, there would be conflicting attributes to the situation that would require resolution.

Initially it is likely that the first coaction fields in HyperReality will be in locations dedicated to specialist activities, such as classrooms, surgeries and shops. By adding the dimension of virtual reality to these specialist places, they will become HyperClasses, HyperResorts, HyperSurgeries and HyperShops. This suggests that the home could develop as a generic HyperReality site and as a protective locus for going to school or the doctor's or the shops as telepresences in HR.

This last point perhaps needs more consideration. One of the main problems with indulging in the fantasies of virtual reality is what to do about the body that is left behind in physical reality as the mind is tricked into thinking that it is flying or fighting in a mythical world. In HyperReality a person by definition is perceptually aware of the physical world around them, yet part of the attention normally given to the physical reality is given to interacting with virtual reality. It is difficult as yet to see how much this matters, but the increasing use of the mobile phone, which is a primitive form of HR, gives us some feel for the issues. People using a mobile phone can walk busy streets and drive cars while talking to someone who is not there. They are locked into a proto-coaction field that excludes the people around them. It can be distracting and irritating to the people around them and we are creating new social habits and structures to deal with this. People now turn their phones off before a meeting and special places are being reserved in waiting rooms and lounges for people who want to use mobile phones. Similarly, there will need to be a process of social adaptation and environmental adjustment to accommodate the advent of HR.

Internet users can of course clip their mobile phones on to their laptops and jack into the Internet. Soon that will mean that they can also click on a desktop icon that can jack them into HyperReality. They will choose a coaction field and, as with such Internet VR programmes as Active Worlds, they will select an avatar to be their telepresence in the virtual dimension of HR. It could resemble them and have a virtual wardrobe from which to dress appropriately, with an interesting selection of accessories. Hair could be grown, cut and styled from a menu or an HR hair designer called in for that special occasion in HR. They could dial up the people they want to join them or, if the meeting was pre-set automatically, suddenly find themselves communicating in HyperReality.

HYPERREALITY AS A TECHNOLOGICAL PARADIGM

(Paradigmatic) terms such as 'artificial intelligence', 'nanotechnology', 'virtual reality' and the 'virtual class', which carry the connotation of a technology that is coming, have a special function. They encourage forward thinking, they evoke futures scenarios, they create a mindset and they act as banners under which regiments of researchers can gather.

(Tiffin 1996)

Not inconsistent with the previous meanings of 'paradigm' is the 'technological paradigm' (Tiffin 1996). The term refers to abstract concepts of particular technologies, their key elements, how they interrelate and function, and their purpose. It is from such paradigmatic knowledge that particular syntagmatic examples can be created. So the Model T Ford, the VW Beetle and the Bugatti Type 57 can be considered as syntagms of the technological paradigm of the motor car. This parallels the paradigmatic/syntagmatic relationship in any code and a basic assumption that there is a dynamic relation between the paradigmatic and syntagmatic aspects of a code. A language changes because of the way it is used. By the same token, a new model of car is created in the light of changes to the car paradigm that have taken place as a result of experience with previous models.

Although the English language used at the beginning of this century had fewer words and some of them were used with different meanings, we can understand books written at that time. Similarly, at the close of the twentieth century cars made at the beginning of the century could be driven along a motorway and were recognisable as cars. It would not, however, be possible to recognise today's PC in the mainframe computers of fifty years ago. The dynamic interaction between syntagm and paradigm in information technology paradigms is faster. In the case of such advanced technological paradigms as artificial intelligence (AI), cloning and nanotechnology, the paradigmatic dimension has outpaced the syntagmatic dimension. The paradigm has become one of speculative possibility in advance of what is syntagmatically feasible. These are paradigms of what could be in the future, rather than of what can be now. HyperReality is a technological paradigm of this kind.

Actual syntagms of HyperReality are the prototype developments at the Advanced Telecommunications Research (ATR) laboratories and current HyperClass experiments at Waseda University (Chapter 1 this volume). The HR paradigm is based on speculation as to how this work could be adapted for use in societies that are in the process of transforming from an industrial base to an information base. In this first stage of development HR is seen as a place-based technology: this is the paradigm that serves for designing the next generation of syntagmatic expressions that, like the mass media of the industrial age, will be accessible in some kind of room. In the early part of the third millennium, however, HR will take on a momentum of its own. It will become independent of location and an everyday extension of our total communication environment, like the mobile phone. Already we are trying to conceptualise generations of paradigms.

As yet we can see no limits to the development of computer processing capability, which means that as yet we see no limits to the development of virtual reality. As yet we have still to find limits, other than the speed of light and the electromagnetic spectrum, to the transmission of information and hence of virtual reality. As yet we have not established the amount of information that can be stored for access at microsecond speeds, so that there seem to be no parameters to the virtual people, the virtual objects and the virtual settings that can be stored for HyperReality. As yet, therefore, we have no conception of what final form, if any, HR will take, and what syntagms may come from it.

Other technological paradigms will impact profoundly on HyperReality. Nanotechnology is the development of molecular machines. Drexler (1990) argued that computer processing and storage would continue to be miniaturised until it reached the molecular level and that, in so doing, computers would, from a human perspective, disappear into the environment. It becomes possible to think of wearing information technology. The fabric of the clothes we wear in the future could form a broadband communications network, with every intersection of the weave a nanocomputer. Stimuli from the immediate environment as well as information from the telecommunications environment could be mediated and matched by clever clothes. We would be suited for HyperReality.

There is a technological drive to make the interface with HR something of which we cease to be conscious. This could be achieved by wearing a complete bodysuit and helmet that totally mediate all proximal stimuli, so that the insertion of virtual elements into our perception of adjacent physical reality could seem natural and normal. It would also mean that our perception of physical reality could be adjusted. We could see clearly in low light conditions or 'see' with parts of the electromagnetic spectrum that are not accessible to the naked eye. We could adjust sound for range and directionality so that it would be possible to hear a distant conversation or the heartbeat of a nearby bird. Our clever clothes could adjust the transmission of heat, texture and physical pressures from the physical environment to our bodies.

Drexler conceptualises a suit made with nanotechnology that would precisely contour a person's body. Like conventional clothes, it would adjust to the movement of a body but, unlike conventional clothes, it would not be supported by the body. It would instead be an autonomous structure that mimicked the movement of the body it contained, and it would have the strength to lend support to that body. Drexler goes further to suggest how such a suit could be a total life-support system. If such a suit is also an HR environment, then there will be no need for rooms and buildings to live in, at least not in physical reality. The HyperWorld we inhabit could well be a palace from the Arabian Nights and we could seem to sleep on a cloud.

Another development that will impact on HR is that of telerobotics. The interaction between what is virtual and what is real is only an interaction of information. Coaction in HR is not trophic. It is possible to see, smell and touch a virtual apple, but not to eat it. A virtual doctor can attend a real patient and undertake a visual and verbal examination, but the virtual doctor cannot perform an operation. Some kind of robotic surrogates will be needed for functions that involve the manipulation of real things by a virtual person.

Artificial intelligence (AI) is another field that will profoundly influence HyperReality development. Although there are questions as to whether computers can ever be intelligent in the way humans are (Searle 1992; Penrose 1989; Edelman 1992), they steadily get smarter in their own way within specific knowledge domains. To add to an appearance of intelligence is the association of artificial intelligence with the development of speech recognition, speech synthesis and image recognition. Computers are beginning to see, hear and talk and are slowly learning and becoming expert at certain tasks. So the artificially intelligent creatures of HyperReality will be able to see, hear and talk as they interact with humans in HR. Virtual people and virtual creatures in HR have their origins in a database and owe their dynamics to computer software. Virtual objects and virtual settings, along with virtual people and virtual creatures, will acquire their own autonomy. They will have an existence independent of a human observer in physical reality. Virtual settings can have their own days and seasons. There can be virtual sunrises and tempests. Virtual footballs and tennis balls can have their own ballistics, size, colour and shape and need not be answerable to the laws of physical reality. Virtual settings and objects could vary in the way they relate to reality. Gradually there will develop a virtual world of virtual people, virtual objects and virtual scenes that exist within the shared memories of computers, and people will learn to live at the intersection of the virtual and the real that is HyperReality. This could come to constitute the basic paradigm by which we know existence in the information society.

PARADIGM SHIFT

The argument over whether the progress of man must proceed in an orderly, continuous extrapolation from the past and present into the future, or whether on the other hand, disjunctures inevitably occur that are not directly predictable from the past goes on.

(Heinich 1970: 53)

Why should a change of paradigm be called a revolution? In the face of the vast and essential differences between political and scientific development, what parallelism can justify the metaphor that finds revolutions in both?

One aspect of the parallelism must already be apparent. Political revolutions are inaugurated by a growing sense, often restricted to a segment of the political community, that existing institutions have ceased adequately to meet the problems posed by an environment that they have in part created. In much the same way, scientific revolutions are inaugurated by a growing sense, again often restricted to a narrow subdivision of the scientific community, that an existing paradigm has ceased to function adequately in the exploration of an aspect of nature to which that paradigm itself had previously led the way.

(Kuhn 1962)

Thomas Kuhn in his book on scientific revolutions used the term 'scientific paradigm' to describe a generic system of scientific thought where the key theories and concepts integrate to form an explanation of the world that is axiomatic in a particular period of time. The term 'technological paradigm' was used by Carlotta Perez (1983) in an adaptation of Kuhn's (1962) concept of scientific paradigms to technology. Heinich (1970) adapted Kuhn's concept of paradigm to education. It became possible to think of science or education or technology as having, over a certain period of time, a paradigmatic set of ideas that are interrelated to form a norm system of thought to which practitioners subscribe and from which they derive their practice.

Kuhn (1962) is writing about revolutions and goes on to suggest that researchers who apply their established science to real-life situations may discover anomalies that cannot be explained within the paradigm of their time. It is the pursuit of an anomaly that leads to the emergence of a new paradigm. Such a shift from an old paradigm to a new paradigm is, Kuhn argues, not without conflict between proponents of rival paradigms.

In Ethiopia in the 1950s there were few motorable highways or modern towns, and most people lived on the land in a subsistence economy. They grew what they ate, made their own clothes and entertainment, built their houses with the help of neighbours and lit them with tallow that came from the fat of their animals. They went to the local market on foot, mule or horse and knew little of modern transport. Their reality was a natural one, derived from the mountains, rivers, forests and fields that they could see and walk among. Since they were largely illiterate and had no television, they had little concept of the world beyond their ken. The places and people in the biblical stories that the local priest read to them had more reality in their minds than the stories they heard of the lands from which foreigners came. At times in the capital city Addis Ababa it was possible to see people from rural areas trying to cross the busy streets. They would crouch in fear, sometimes threatening passing cars with their spears, then make a dash into the road only to turn back in terror from an oncoming car. Sometimes they would appear at the entrances to large stores gazing in awe at the produce on the shelves. These were people who lived within the paradigm of a pre-industrial society. We can only imagine the trauma they experienced in coming to the man-made artificial reality of a modern city and their difficulty in understanding the ways things were done and the rules by which things happened, as they underwent a paradigm shift in their sense of what constituted reality.

The generic 'we' in the previous sentence refers to the people who, like me the author and you the reader, can read. Reading is one of the basic skills of those who live in urbanised societies. The skill confers the ability to conceive of reality in the abstract, to know worlds beyond the direct experience of the individual and to contribute to the creation of the Euclidean realities in which 'we' live. Literacy has been the prime concern of national education systems that, along with a predominantly urban way of life, characterise modern industrial societies. Mass education enables citizens to share a common belief in realities that are essentially abstract concepts. People kill and die for their nation, religion, culture and political creed and put their trust in governments, banks and taxes to transform what they do into what they want. And what they want is based on utopian notions of ideal realities that are always receding. The transactional economies of industrial societies depend on shared world views of the way things work and a belief in scientific explanations of why they work. 'We' are born and brought up in, and most of us are essentially convinced of, the paradigm of reality of the industrial society. If we were not, it would not work.

Islands of urbanised societies have been around for thousands of years, but it is only in the last hundred years or so that the paradigm of the industrial society has become the dominant world-view. Even societies that do not have the physical reality of an industrial society, subscribe to the vision of one and strive to realise it. But at the very point that the paradigm of the industrial society becomes global, it falls into question. As it does so, it begins to unravel.

Heilbroner (1995) looks at a pre-industrial time when people expected that the future would be much the same as the present and past. He argues that industrial societies were characterised by a belief that the future was something to work for because it would be better than the past. However, he finds today a new spirit of pessimism in post-industrial societies that see the future as threatening. Characteristic of such thinking is that of the postmodern movement in western intellectual circles, which profoundly criticises the scientific and technological paradigm of the industrial society and the technological assumptions inherent in the concept of an information society. Umberto Eco (1986) and Jean Baudrillard (1988) use the term *hyperreality* to describe the way our perception of the world is increasingly dependent on simulations of reality, as we become an information society. In the mass media of America, and in institutions such as Disneyland, they see a tendency for signs to break loose from their referential moorings, to fly free of cognitive meaning and take on a hyperlife of their own that is more real than reality and hence *hyperreal*.

Nobuyoshi Terashima coined the term HyperReality to define a technological metaconcept. It is written as a single word but capitalised twice to allow the abbreviation HR. This in turn can be readily matched with the widely accepted abbreviation of VR for Virtual Reality. It also distinguishes the term from the postmodern use of *hyperreality*. Yet there is a link.

Terashima derived the metatechnological concept of HyperReality from his work in leading a research team that was trying to solve a fundamental problem of communication between the real and the virtual. He was trying to find a technological solution to what postmodernists perceive as a problem. In HyperReality he has created a technology that enables *hyperreality*. The coincidental resemblance of the terms seems appropriate.

'Hyper' means an extra dimension beyond the normal. HyperReality means a reality in which there is the extra dimension of virtual reality within normal physical reality. But this, as we have seen, is not a simple add-on of another set of capabilities. For the human species it will be a fundamental reformulation of their perception of reality and of the world they live in.

REFERENCES

Baudrillard, J. (1988) *America*, New York: Verso.

Bell, D. (1976) *The Coming of Post-Industrial Society: A Venture in Social Forecasting*, New York: Basic Books.

Beniger, J.R. (1986) *The Control Revolution: Technological and Economic Origins of the Information Society*, Cambridge, MA: Harvard University Press.

Bretz, R. (1953) *Techniques of Television Production*, New York: McGraw-Hill.

Castells, M. (1997) *The Power of Identity*, Oxford: Blackwell.

Dordick, H. and Wang, G. (1993) *The Information Society: A Retrospective View*, Newberry Park, CA: Sage.

Drexler, E.K. (1990) *Engines of Creation*, New York: John Wiley.

Eco, U. (1986) *Travels in Hyperreality*, London: Pan.

Edelman, G.M. (1992) *Bright Air, Brilliant Fire: On the Matter of the Mind*, New York: Basic Books.

Heilbroner, R. (1995) *Visions of the Future*, Oxford: Oxford University Press.

Heinich, R. (1970) *Technology and the Management of Instruction, Monograph 4*, Washington DC: Association for Educational Communications and Technology.

Jayaweera, N.D. (1987) 'Communications satellites: a Third World perspective', in F. Finnegan, G. Salomon and K. Thompson (eds), *Information Technology: Social Issues*, Sevenoaks: Hodder and Stoughton/Open University Press.

Kranzberg, M. and Pursell, C.W. (1967) *Technology in Western Civilisation*, New York: Oxford University Press.

Kuhn, T. (1962) *The Structure of Scientific Revolutions*, Chicago: University of Chicago Press.

Penrose, R. (1989) *The Emperor's New Mind: Concerning Computers, Minds and Laws of Physics*, New York: Oxford University Press.

Perez, C. (1983) 'Structural change and the assimilation of new technologies in the economic and social systems', *Futures*, 15: 357–375.

Saussure, F. de (1916) *Course in General Linguistics*, London: Fontana/Collins.

Searle, J.R. (1992) *The Rediscovery of the Mind*, Cambridge, MA and London: Bradford.

Tiffin, J. (1996) 'The Virtual Class is coming', *Education and Information Technologies*, 1, 2.

Toffler, A. (1970) *Future Shock*, New York: Bantam Books.

Winner, L. (1977) *Autonomous Technology*, Cambridge, MA: MIT Press.

Watson, J. (1993) A *Dictionary of Communications and Media Studies*, third edn, London: Edward Arnold.

3

HYPERVISION

Narendra Ahuja and Sanghoon Sull

Editors' introduction
The challenge at the heart of HyperReality is the idea of teleporting together three-dimensional images of people, things and places that are distant from each other. We teleported sound when Bell invented the telephone. We teleported two-dimensional images when Baird invented television. Now we can do it with three-dimensional images. This means that if you have a camera on a crossroads in London, then you can observe the traffic patterns in 3D from an office in Tokyo and discuss the traffic flow problem with a group of experts in both places.

In this chapter Ahuja and Sull describe how two-dimensional images can be converted into three-dimensional images and their own contribution to this work. This is not an easy read, but then it is not an easy subject. We are reminded of the problems faced in cartography in seeking to show a three-dimensional world on a two-dimensional page. A camera is a device that does this automatically. Ahuja and Sull start with the two-dimensional picture of reality and seek to create from it the third dimension. This is a bit like working from an atlas to reconstruct a three-dimensional globe.

INTRODUCTION

HyperWorld is an advanced world where images from the real world are systematically integrated with synthesised images. One of the key elements necessary to implement this HyperWorld is the technology that enables realistic visual communication of virtual three-dimensional (3D) objects to observers in the HyperWorld. Such visual communication can be achieved using computer vision techniques by first estimating the 3D structure of an object from still or video images taken from a sensor such as a camera (analysis), and then by synthesising the different views. Using the camera analysis results, the synthesis can be done in two ways: (1) by showing some selected characteristics of the 3D scene, or (2) by displaying the original or enhanced photometric appearance of 3D objects from an observer's viewpoint. In this chapter, two approaches for such analysis-guided synthesis will be described.

43

First, if the goal of the synthesis is to communicate to the observer certain characteristics of the scene, which are relevant to its purpose (such as those that may be useful for navigation), then a feature-based approach, such as a line sketch, may suffice for a very low bit-rate visual communication. This approach uses the notion that diverse image features such as points, lines and regions that have a locally compatible 3D rigid motion and structure would contribute to realistic synthesis, thus suggesting an approach to analysis-guided video coding. It is not expected that synthesis will retain the original photometric appearance of the images pixel by pixel, but the approach could be combined with conventional coding techniques to achieve such characteristic visual fidelity. Second, if an object can be represented by a model, the analysis and synthesis process becomes easier and more robust.

Two examples of synthesis will be described. Both involve the use of aerial video sequences taken over relatively flat ground. The first, more general, example used video of desert and does not involve specific object models, whereas the second shows an airport and uses runway models for the analysis and synthesis. Using a detailed planar model of runways, the runway scene can also be augmented or synthesised by supplying missing features to the input video, e.g. by reconstructing occluded parts and emphasising selected features. Before presenting the details of the two approaches, a brief summary of the general field of 3D vision is given in the following section.

3D COMPUTER VISION

Computer vision is concerned with computational understanding and use of the information present in visual images. The methods discussed in this chapter are primarily aimed at 3D interpretation of image content. Following are some representative areas and problems in 3D computer vision.

Sensing and image formation

Computer vision begins with the acquisition of images. A camera produces a grid of samples of light received from different directions in a scene. The position within the grid where a scene point is imaged is determined by perspective transformation. The amount of light recorded by the sensor from a certain point in a scene depends upon the type of lighting, reflection characteristics, the orientation of the surface being imaged and the location and spectral sensitivity of the sensor.

Image segmentation

The objective of this stage is to compress the huge amount of image detail by identifying and representing those aspects of image structure that are salient for later stages of interpretation. Typically, this is accomplished by detecting homogenous regions in the image or on its edges. This must be done for all degrees and sizes of region homogeneity, yielding a multi-scale segmentation. Further, small regions may group together to form a larger structure seen as a homogenous texture that may define another basis for characterising the homogeneity of a segment. The result of segmentation is a partitioning of the image such that each part is homogenous in some salient property relative to its surround.

3D interpretation

One central objective of image interpretation is to infer the 3D structure of the scene from images that are only two-dimensional (2D). The missing third dimension necessitates that assumptions be made about the scene so that the image information can be extrapolated into a 3D description. To this end, the presence in the image of a variety of 3D 'cues' is exploited. These cues may occur in:

1   a single image taken by a camera at a single time instant,
2   a sequence of images acquired by one camera over a time interval (this is the focus of the work described later in this chapter),
3   a set of images taken by multiple cameras at a single time instant,
4   a time sequence of images taken by multiple cameras.

Examples of the different kinds of cues and their significance are given below. In each case, the main task is to devise algorithms that can estimate 3D structural parameters from image-based measurements.

Image curves

The boundaries of image regions, or the curves comprising a line drawing image, reveal scene characteristics such as the extent and shape of an object, and occlusions between objects.

Shading

The nature of gradual spatial variation of image values within a region is related to surface shape characteristics such as whether it is convex or concave, planar or curved.

Texture gradient

Variation in the coarseness of image texture is indicative of how the corresponding texture surface is oriented in the scene. For example, in an image of a large field of flowers, the more distant flowers would be packed more densely together than the flowers in the foreground.

Stereo

Suppose two cameras are placed at different locations with different orientations, like human eyes. In this case, the coordinates of the projections of a given point in the scene are different and their disparity is trigonometrically related to the 3D position of the point in the scene. Such disparity variation within an image can be used to estimate the 3D surfaces in the scene. Three or more cameras can be used instead of two.

Dynamic cues

If there is relative motion between the scene and the camera(s), then the image data consists of dynamic image sequence(s). Each of the above-mentioned cues then provides additional information, since observations about the temporal behaviour of the cue are also available, in addition to its spatial properties. For example, the temporal variation of image values, called optical flow, can be used to estimate the relative motion, surface shape and layout of objects. A moving object gives rise to a decreasing amount of flow as its distance from the camera increases. Similarly, the curves of a moving image, such as those corresponding to orientation discontinuities on the surface of an object, or the silhouette of a rotating object, yield a wealth of information about the object's shape as well as about how it is moving. Sequences may be taken by multiple cameras simultaneously and this makes surface estimation easier because a larger amount of data is available from the spatial disparity as well as its variation over time.

Active image acquisition

If the cameras maintain a single, fixed geometrical or optical configuration, the amount of additional information extracted from successive images depends on changes occurring in the scene, for example due to relative motion. It is possible to maximise the increment in scene information obtained from successive images by dynamically reconfiguring the cameras so that the cues in the new image are more informative. Thus, a partial interpretation is used to dynamically control the sensing parameters so that each stage of image acquisition adds the maximum possible information to the interpretation. For example, if a scene is too bright, the aperture size may be reduced; if the object is not in sharp focus, the focus setting may be changed; and if the object is not well placed with respect to both cameras in stereo analysis, the cameras may be moved to fixate on the object. All of these happen in parallel. As a certain part of the scene is satisfactorily interpreted, the results of the interpretation may be used to determine where to point the cameras next and even to suggest how to analyse the new parts so as to obtain the best final interpretation in the minimum time.

Representation

The 2D structure of an image or the 3D structure of a scene must be represented so that the structural properties required for various tasks are easily accessible. For example, the hierarchical 2D structure of an image may be represented through a pyramidal data structure that records the recursive embedding of the image regions at different scales. Each region's shape and homogeneity characteristics may themselves be suitably coded. Alternatively, the image may be recursively split into parts in some fixed way (e.g. into quadrants) until each part is homogenous. This leads to a tree data structure. Analogous to 2D, 3D structures estimated from the image-based cues may be used to define 3D representation. The shape of a 3D space or object may be represented by its 3D axis and how the cross section about the axis changes along the axis. Analogous to the 2D case, the 3D space may also be recursively divided into octants to obtain a tree description of the occupancy of space by objects.

ANALYSIS-GUIDED VIDEO SYNTHESIS

Video synthesis is one of the key components of communication between observers in a HyperWorld. The method for augmentation and synthesis of images and video data may be described as first extracting the 3D content of the data and then using the 3D information to produce the results. Alternatively, this amounts to coding the data in terms of higher-level attributes than those used in familiar coding methods. Existing methods for video coding can be divided into three broad categories:

1   those based on 2D signal coding,
2   those based on 3D models,
3   those based on image features.

  Methods based on 2D signalling encode the video by using conventional signal processing and communication techniques. MPEG-1 coding is based on DCT (Discrete Cosine Transform) and motion compensation. These methods preserve the original photometric values of images as much as

NARENDRA AHUJA AND SANGHOON SULL

possible and therefore the typical compression ratio is not high. The methods in the second category are based on 3D models. For example, if a 3D model (e.g. the human face models proposed in Aizawa and Harashima 1989 and Terzopolous and Waters 1993) is available, moving pictures are analysed to extract the model parameters and only the model parameters are transmitted. Then, at a receiver, the original face motions are synthesised by deforming the face model using the model parameters. Although the compression ratio of these methods is very high (only model parameters need to be sent), it only works for the specific type of images that a 3D model holds and it is sometimes difficult to automatically extract the model parameters. However, once the model parameters are estimated, an arbitrary viewpoint of the object can be synthesised. Another example of a 3D model is the planar surface runway model mentioned earlier. The model is described later in the paper and shows how a runway video is synthesised.

Two approaches for video coding or synthesis are described. First, we describe a feature-based synthesis approach with very low bit-rate visual communication. Second, we present an example of 3D analysis-guided video synthesis that synthesises and augments input video by adding missing features, e.g. by reconstructing occluded parts and emphasising selected features under the assumption that the object in the scene can be modelled as a planar runway. Experiments with feature-based synthesis of three video sequences are reported in which the image sizes are compressed by 810, 1,004 and 734, but in which the synthesised sequences appear compellingly similar to the originals when the two are played side by side on a monitor. The compression ratio is defined as the ratio of the memory size for the original data to the size for the compressed data.

Feature-based synthesis

A feature-based approach uses image features to depict the salient motion (similar to line sketch) of the input video (Sull and Ahuja 1994; 1995). Although this approach requires a 3D analysis when the synthesis is to be done from a viewpoint different from the original video, it does not need difficult 3D analysis when the same view is needed. A key feature of this approach is an integrated use of diverse image features. Diverse features are detected using existing feature detection methods (Sethi and Jain 1987; Debrunner and Ahuja 1990; Liu and Huang 1991; Tabb and Ahuja 1993), then we use a systematic way of selecting those features that preserve the characteristics of 3D motion of an observer relative to a scene. Diverse features are matched and segmented so that all features in each segment have the same 3D motion from a pair of 2D successive images. These matched features provide an efficient means of synthesis. If all the extracted features are simply displayed, observers may perceive false motion. This is because feature detectors are sensitive to noise, such as slight illumination changes in the image capturing process and, consequently, they sometimes cannot detect corresponding features in two images. Diverse features carry information about motion and structure to different degrees and have different, often complementary, strengths and shortcomings. Thus, when a given feature does not contribute significantly to the matching and segmentation process, other more pertinent features help achieve reliable correspondence. This list of features could be changed to achieve better perceptual synthesis while still following the basic approach presented. Point, line and region features are used in the current algorithm, although other types of feature can easily be incorporated into the paradigm.

The method for integrated matching and segmentation is based on local affine approximation of the displacement field, which is derived under the assumption of locally rigid motion. Thus, the displacement vector $[d_x, d_y]'$ at t_1, which represents the displacement caused by motion of a point located at (x, y) is locally approximated by:

$$d_x = c_0 + c_1 x + c_2 y$$

$$d_y = c_3 + c_4 x + c_5 y$$

Each distinct motion is represented in the image plane by a distinct set of values of six parameters $c = \{c_0, c_1, c_2, c_3, c_4, c_5\}$. All sets of values supported by feature locations in two adjacent frames are identified by exhaustive coarse-to-fine search. However, to reduce computational complexity the 6D parameter space is decomposed into two disjoint 3D spaces. The support for any set is computed from the image plane distances between the observed feature locations and those predicted by the parameter values. The well-supported sets of values thus found yield two results simultaneously. First, the established correspondences between features, and second, they segment the features into subsets corresponding to locally rigid patches of the moving objects. Since the matching of features is based on 3D motion constraints, problems due to motion, object boundaries or occlusion can be avoided. Further, our method can handle large motion as well as small motion. The integrated use of diverse features not only gives a larger number of features (an overconstrained system) but also reduces the number of candidate matches for features thus making the matching less ambiguous. Feature-based video coding needs to send information on feature locations and affine coefficients. In this chapter the term point feature denotes both a distinguishing image point as well as the intersection of lines (line points). The former is defined as a point whose location corresponds to the local maxima or minima of the intensity values. Three examples of such synthesis are described below.

Indoor scene

For a features-based synthesis for an indoor scene, we obtained six segments, each of which corresponds to a part of the scene with similar depths. This is the result of integrated interpretation. These are diverse but mutually compatible features. Note that this synthesis is done from the same viewpoint as the/input. The compression ratio achieved is 810.

Desert scene

We derived a sequence from a commercially available VHS video tape of a film shot from a flying aircraft. The original sequence and the resulting synthesised sequence based on feature-based synthesis incorporated vanishing lines, obtained using the 3D motion and structure analysis method (Sull and Ahuja 1994). A vanishing line is defined as the intersection of the image plane with a plane that includes the camera centre and is parallel to the object plane. If we watch the synthesised video as it is played on a monitor, we perceive the same motion and structure in an informal viewing as from the original image sequence. Note that the synthesised video can be shown in binocular (stereo) display, as well as from a viewpoint different from the input, since the necessary 3D information is available from the analysis. Binocular display further highlights the recovered motion and structure. The compression ratio achieved is 1,004.

Runway scene

We derived a sequence from a commercially available CAV laserdisc of film shot from a flying aircraft. This is a challenging sequence since the images contain partially or completely occluded vanishing lines and there is reflection of the ground on the bottom of the aeroplane. The quality of the image is rather better than the desert sequence images obtained from VHS tape.

3D model-based synthesis

The 3D motion and structure estimation process enables not only video synthesis from an arbitrary viewpoint, but also a variety of enhancements of an input sequence. We describe three examples of video enhancement for a flight sequence containing a runway:

1 colouring the part of the scene that is above a vanishing line, say blue, to show sky,
2 filling in the parts of the scene that are occluded,
3 repainting the runway edges and placing yellow discs along them to simulate lights.

Such enhancements are facilitated by the 3D analysis of the image sequence and in this sense represent a higher level approach to video enhancement than the classic 2D signal processing methods. The exact nature of enhancements is domain-dependent (as expected), but they are all described in 3D terms, and their implementations are therefore based on the results of the 3D analysis stage. Details of the algorithm can be found in Sull (1993).

Automatic colouring of selected scene parts

The 3D analysis process produces an estimate of ground orientation that is equivalent to the estimated vanishing line. Therefore, the part of the image above the estimated vanishing line can be coloured blue, assuming that it corresponds to the sky. Without using the output of 3D analysis, it would be very tedious to colour it manually. It would in fact be impossible when the vanishing line is occluded.

Reconstruction of occluded scene parts

We can reconstruct any part of the ground that is occluded in some frames, for example by an aeroplane or a moving vehicle, if that part appears in other frames. The reconstruction is performed by using the estimated displacement field obtained from the 3D analysis stage. First, the occluded parts of the images are identified. This can be done by registering two images according to the given displacement field. Irani and Peleg (1992) provide an example of the reconstruction of occlusions. In our case, since the ground is partially occluded in some frames by the bottom part of the aircraft, which occupies the same portion of the image plane across the sequence of images, the fixed upper rectangular portions of each image are removed manually for simplicity. Then, by using the given displacement field, the occluded parts are reconstructed by interpolating the intensity values of the image locations corresponding to the sections of ground that are visible in either direction in the time domain.

Enhancement of recognised runway

If the parts of the image corresponding to runway edges can be identified, they can be enhanced so that they can be more clearly seen. An algorithm that identifies runway edges from extracted line features in input images is now briefly described. The algorithm consists of four steps:

- The first step is to extract line features from input images.
- The second step is to compute the direction of flight, based on the motion estimates obtained from the 3D analysis stage.

- The third step is to find correspondences between the 3D runway model lines and 2D line segments in each frame. In this case, to reduce the search space for finding correspondences, we use the domain-specific knowledge that the direction of a runway is approximately parallel to the direction of a moving camera as it projects on to the ground.
- The fourth step involves non-iterative computation of the model parameters from the candidate lines whose correspondences were found in the third step.

The original sequence and the resulting synthesised sequence derived from feature-based synthesis incorporated the vanishing lines obtained by the 3D motion and structure analysis method (Sull and Ahuja 1994). If we watch the synthesised videos as they are played on a monitor, we perceive the same motion and structure from them in an informal viewing as from the original image sequence. The compression ratio achieved is 734.

The original sequence is enhanced by adding the results of 3D analysis. For the reconstruction of occluded parts, the fixed portion of the image plane, size 640 by 125, was manually deleted from the top. The occluded parts are reconstructed using bilinear interpolation. A circle pattern along the runway edges is painted, using the super sampling ray tracing technique (Foley *et al.* 1992). In this example, the occluded parts of the earlier images were generated from information taken from later images in the sequence. Note that the synthesised video can be shown in binocular (stereo) display, as well as from a viewpoint different from the input, since the necessary 3D information is available from the analysis. Binocular display further highlights the recovered motion and structure parameters.

REFERENCES

Aizawa, K. and Harashima, H. (1989) 'Model-based analysis image coding system for a person's face', *Signal Processing: Image Communication*, 1, 2: 139–152.

Debrunner, C. and Ahuja, N. (1990) 'A hankel matrix based motion estimation algorithm', *Proceedings International Conference Computer Vision Pattern Recognition* (Atlantic City): 384–389.

Foley, J.D., van Dam, A., Feiner, S.K. and Hughes, J.F. (1992) *Computer Graphics: Principles and Practice*, second edn, New York: Addison Wesley.

Irani, M. and Peleg, S. (1992) 'Image sequence enhancement using multiple motions analysis', *Proceedings IEEE Conference Computer Vision Pattern Recognition*: 216–221.

Liu, Y. and Huang, T. (1991) 'Determining straight line correspondences from intensity images', *Pattern Recognition*, 24: 489–504.

Sethi, I. and Jain, R. (1987) 'Finding trajectories of feature points in a monocular image sequence', *IEEE Trans. Patt. Anal. Mach. Intell.*, PAMI-9: 56–73.

Sull, S. (1993) 'Integrated 3D analysis and analysis-guided synthesis', unpublished Ph.D. thesis, University of Illinois, Urbana-Champaign.

Sull, S. and Ahuja, N. (1994) 'Integrated 3D analysis and analysis-guided synthesis of flight image sequences', *IEEE Trans. Patt. Anal. Mach. Intell.*, 16: 357–372.

Sull, S. and Ahuja, N. (1995) 'Integrated matching and segmentation of multiple features in two views', *Computer Vision and Image Understanding*, 62: 279–297.

Tabb, M. and Ahuja, N. (1993) 'Detection and representation of multiscale low-level image structure using a new transform', *Proceedings Asian Conference Computer Vision* (Osaka, Japan): 155–159.

Terzopolous, D. and Waters, K. (1993) 'Analysis and synthesis of facial image sequences using physical and anatomical models', *IEEE Trans. Patt. Anal. Mach. Intell.*, 15: 569–579.

4

VIRTUAL HUMANS

Nadia Magnenat-Thalmann, Prem Kalra, Laurent Moccozet

Editors' introduction
Nobuyoshi Terashima sees virtual inhabitants of HyperReality as being derived either from camera shots of people transmuted into 3D in the manner described by Ahuja and Sull, or as life forms created with computers. Nadia Magnenat-Thalmann's research has been dedicated to the development of computer-generated autonomous virtual people. With her colleagues, Prem Kalra and Laurent Moccozet, she describes the history and state of the art of the development of 3D virtual humans. They focus on the particular approach taken at MIRALab for creating and animating realistic virtual humans for real-time applications. They detail the complete animation framework, then go on to present a case study: CyberDance. This application combines high technology with art choreography in a dance. In this performance, the movements of the dancers are captured and replicated, in real time, by their virtual counterparts. Many people feel Gothic overtones in the idea of virtual humans but, as the CyberDance case study shows, this does not have to be horrific. Nadia Magnenat-Thalmann imagines a time in the future when her great-great-great grandchildren will lie tucked up in bed and need only whisper her name for a virtual version of herself to emerge from the shadows, sit on their bedside and with infinite patience tell story after story.

INTRODUCTION

Virtual humans are computer models of people that can be used as representations of ourselves or other live participants in virtual environments and in HyperReality. Computer modelling of virtual humans includes research in several domains such as representation of shapes and motion, biomedical simulation of complex structures, verbal and non-verbal communication, human factor analysis, and real-time issues for inhabiting virtual environments with autonomous virtual humans. These research areas enable diverse applications of virtual humans in engineering and design, education and training, games, TV and films, and medicine. Virtual humans are now the foundation of applications requiring personal and live participation.

The first computerised models were created more than twenty years ago. The main idea was to simulate a very simple articulated structure for the study of problems of ergonomics by airplane and car manufacturers. In the 1970s, researchers developed methods of animating human skeletons, mainly based on interpolation. *The Juggler*, from Information International Inc. (1982), contained the first somewhat realistic human character in computer animation. However, the human shape was completely digitised and body motion was generated using 3D rotoscopy with no facial animation. The first 3D procedural model of human animation was used in *Dreamflight* (Magnenat-Thalmann, Bergeron and Thalmann 1982), a 12-minute film, one of the first to feature a 3D virtual human. In the 1980s researchers started to base animation on keyframe animation, parametric animation, and, later, on the laws of physics. Dynamic simulation made it possible to generate complex motions with realism. However, the main complexity in the animation remains the problem of integrating many techniques. True virtual humans should be able to walk, talk, grasp objects, show emotions, and communicate with the environment.

With the latest innovations in interactive digital television (Magnenat-Thalmann and Thalmann 1995), multimedia and game products, a need for systems that provide designers with the capability for immersing real-time simulated humans has emerged. The ability to place the viewer in a dramatic situation created by the behaviour of other, simulated, digital actors will add a new dimension to existing simulation-based products for education and entertainment on interactive TV. In the games market, convincing simulated humans have been identified by the industry as a way of giving a fresh appearance to existing games and enabling new kinds of game to be produced. Finally, in Virtual Reality, representing participants by a virtual actor is an important factor for presence. This becomes even more important in multi-user environments, where effective interaction among participants is a contributing factor to presence (Slater and Usoh 1994). This is self-representation in the Virtual World. Even with limited sensor information, a virtual human frame can be constructed in the Virtual World that reflects the activities of a real body.

We have been working on the simulation of virtual humans for several years. Until recently, these constructs could not act in real time. Today, however, it is important for many applications to simulate virtual humans in real time that look believable and realistic. We have invested considerable effort in developing and integrating several modules into a system capable of animating humans in real-time situations. This includes the development of interactive modules for building realistic individuals and a texture-fitting method suitable for all parts of the head and body. Animation of the body, including the hands and their deformations, is the key aspect of our system; to our knowledge, no competing system integrates all these functions. Facial animation has also been included, as we will demonstrate later in our case study.

We have developed a single system containing all the modules for simulating real-time virtual humans in distant virtual environments.[1] Our system allows us to rapidly clone any individual and to animate the clone in various contexts. There is still no way that people can mistake our virtual humans for real ones, but we think they are recognisable and realistic, as we will show in our case study described later in this chapter. The whole process of animation may be broadly divided into three modules: modelling, motion control and deformation. Modelling though a pre-processing step has a great impact on real-time performance in animation. The motion control module drives the deformation controller. At present there is no reverse feedback from the deformation controller to the motion control. Figure 4.1 shows the logical units of the system.

In this chapter we first describe construction methods for our virtual humans intended for real-time animation. We then go on to describe the animation of the body, animation of the hands and facial animation. This is followed by a section on the animation framework, which integrates all these modules. Finally we present a case study of CyberDance, which uses our animation system.

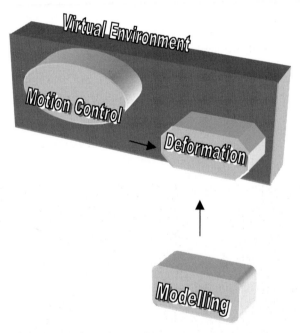

Figure 4.1 Main modules of the system for simulating real-time virtual humans in distant virtual environments.

MODELLING OF VIRTUAL HUMANS

Real-time animation requires a small number of polygons and specific data structures to accelerate the computing process. Due to differences in modelling and animation for face, hands, and body, our virtual humans are divided into separate parts.

Face modelling

For face we have two approaches: a sculpting approach and model fitting using photographs. In the sculpting approach, we use the software *Sculptor* (Magnenat-Thalmann and Kalra 1995), dedicated to the modelling of 3D objects. This approach is based on local and global geometric deformations. Adding, deleting, modifying, and assembling triangle meshes are the basic features provided by *Sculptor*. Real-time deformations and manipulation of the surface give the designers the same facilities as with real clay or wax sculpting.

Using *Sculptor*, instead of a scanning device, allows us to take into account the constraints of real-time animation while modelling the head. Real-time manipulation of objects requires as small a number of polygons as possible, while more polygons are required for the beautification and animation of the 3D shape. Designing the face with this program allows us to directly create an object with the appropriate number of polygons. This can be done knowing which region has to be animated (this requires many polygons) and which region requires less, or no animation at all (this needs fewer polygons). Designers are also able to model simpler objects knowing the texture will add specific details, like wrinkles and shadows, to the object.

Starting from a prototype head accelerates the creation process. Figure 4.2 shows the modelling of a face started from an already existing one.

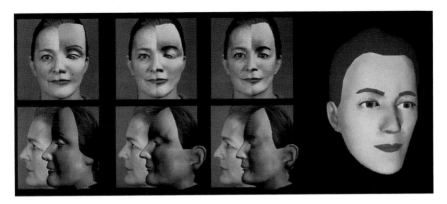

Figure 4.2 Face modelling using the sculpting approach.

Therefore, the more prototypical heads we have, the less time we need to spend. Starting from scratch, the designer can model half of the head, and use a symmetric copy for the other half. In the final stages, however, small changes should be made on the whole face because asymmetric faces look more realistic.

The modelling of the face is done in such a way that we not only construct the shape of the head but also add structure information. This allows the face model to be ready for animation. The structure contains regions, defined as groups of triangles, which determine features and parts of the face that would move during animation. Our face model, unlike many face models from neck to forehead, includes features such as the eyes, tongue, teeth, ears and hair.

In another approach (Lee, Kalra and Magnenat-Thalmann 1997) to constructing a 3D face, we use two orthogonal photographs. In this reconstruction, only some points (so called feature points) are extracted in a semi-automatic manner. These are the most characteristic points by which people are recognised and they can be detected from front and side views. It is a simple approach and does not require high-cost equipment. The extracted features are then used as constraints to the deformation of a canonical 3D face. The deformation model, called DFFD, is used and is described later in the section on the basic hand multilayer model where we use the model for simulating the muscular layer in hand animation. The overall flow for the reconstruction of a 3D face is shown in Figure 4.3. The method can also automatically fit a texture image on to the modified 3D model.

Hands modelling

We use the sculpting approach, as described above for face modelling, also for the modelling of hands. Two basic sets of 3D hands are used, one for

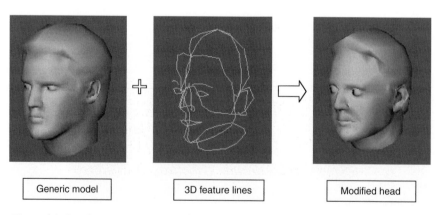

| Generic model | 3D feature lines | Modified head |

Figure 4.3 Semi-automatic human face modelling based on features extraction.

each gender. The existing male hands were refined to give a feminine look using the *Sculptor* program. The proposed hand simulation model (described in the section on the basic hand multilayer model) allows us to model morphological variations. Deformations of the muscular layer can be parametrised according to some morphological changes, such as hand thickness, or skeleton finger length. The muscle layer is first fitted to the skeleton, or scaled to the given morphological parameters, and the muscular operators are then applied to the hand surface to create a new hand. The resulting new hand data set can then be directly used for animation.

Figure 4.4 shows some morphological variations of hands: (a) and (b) show local modification changes. In (a) the length of the middle finger is modified, and in (b), the thumb is extended from the rest of the hand and its size is modified. These two morphological changes are parametrised by the changes of the underlying skeleton. In (c) a global morphology change is shown: the thickness of the hand is increased. This change is parametrised independently from the skeleton.

Body modelling

BodyBuilder is a software package for the creation of human body envelopes developed at LIG-EPFL with MIRALab's contribution to the interface design and testing. The goal was to develop realistic and efficient human modelling that enabled us to use this data for real-time motion and deformation.

A multilayered approach is used for the design of human bodies (Shen and Thalmann 1995):

• The first layer is an underlying articulated skeleton hierarchy. This skeleton is schematically similar to a real human skeleton. All the

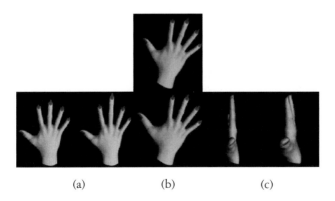

(a)   (b)   (c)

Figure 4.4 Various morphological hand variations.

human postures can be defined using this skeleton. The proportions of the virtual human are designed at this stage.

- The second layer is composed of grouped volume primitives. Volume primitives fall into two categories: blendable and unblendable. These volume primitives are also known as metaballs or ellipsoids. They have different colours attributed to them, depending on the selected function: blendable/unblendable, deformable/non-deformable, or shape: positive/negative. Because metaballs can be joined smoothly and gradually, they give shape to realistic organic-looking creations, suitable for modelling human bodies. They are attached to the proximal joint of the skeleton and can be transformed and deformed interactively. In this way, the designer can define the 3D shape of a virtual human. Figure 4.5 shows a male and female human body prototype created and constructed by positioning, scaling and rotating these volumes as well as by attaching them to a desired joint articulation of the avatar's skeleton. Designers can start from a simplified structure leading to a rough shape, and then go back into the model by increasing the number of volumes. This will give, step by step, a higher level of detail. Two main panels are available as editing tools of the primitives. One includes the creation parameters and various options such as delete/add, and the attachment of the primitives to the skeleton. It also travels through the hierarchy of the model of the skeleton. The other panel was conceived to give control over deformations of these volumes when the prototype is animated, meaning that these can change shape and position/rotation when a joint articulation is activated. This technique is restricted to non real-time application because of the amount of computing technology required, but the realism of the results can be quite impressive. The human form is a very complex shape to reproduce and modelling it is a tedious task, since the human eye is very sensitive to inaccuracies of the human figure. The most challenging part is to manipulate these volumes in a 3D environment in order to simulate the shapes and behaviours of muscles. This requires strong skills in anatomy or in drawing and sculpting human figures.

- The third layer is the equivalent of the human skin. The envelope of the body is created with spline surfaces using a ray-casting method. In this way, metaballs have observable effects on the surface shape. This approach is based on the fact that human limbs exhibit a cylindrical topology and the underlying skeleton provides a natural centric axis upon which a number of cross-sections can be defined.

Texture fitting

Texture mapping is a well-known method in computer graphics for improving the quality of virtual objects by applying real images to them.

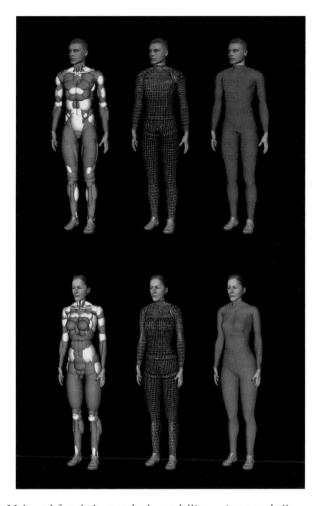

Figure 4.5 Male and female human body modelling using metaballs.

It is a low-cost methodology in terms of computation time, which is very useful for real-time applications. For virtual humans, the texture can add a grain to the skin, including colour details like those for the hair and mouth. These features require correlation between the image and the 3D object. A simple projection is not always sufficient to realise this correlation: the object, which is designed by hand, can be slightly different from the real image. Therefore an interactive fitting of the texture is required (Litwinowicz and Miller 1994). In Figure 4.6, the wrinkles of the hands have been fitted to the morphology of our 3D model: (a) shows the texture which has been applied to the 3D hand shown in (b).

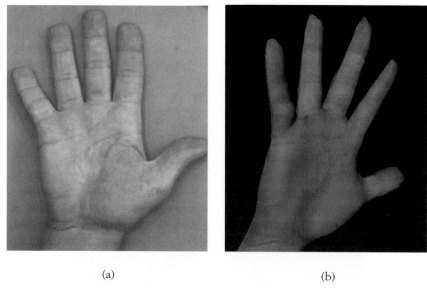

(a)   (b)

Figure 4.6 Texture applied to a 3D hand.

We developed a new program for fitting the texture to the features of a 3D object (Sannier and Magnenat-Thalmann 1997). This enables the designer to interactively select a few 3D points on the object. These 3D points are then projected on to the 2D image. The projection can be chosen and set interactively, hence the designer is able to adjust these projected points to their correct position on the image. This way, we obtain the texture coordinates for the selected 3D points. The problem is that we want to avoid the interactive specification of the texture coordinates for all the 3D points. To map a whole surface, all the 3D points of this surface need their own texture coordinates. We implemented a method for finding these texture coordinates by interpolating from the already existing ones. Thus, all the 3D points are projected on to the 2D image. Using a Delaunay triangulation with the 2D marked points, we can select which points are projected inside the Delaunay area. We consider these points as belonging to the 3D surface to be textured. The barycentric coordinates are calculated for all these projected points inside a Delaunay triangle. This gives the position of each point. After the motion of a marked point (vertex of a Delaunay triangle), the positions of the projected points are recalculated using their barycentric coordinates. Finally, the texture coordinates of all the 3D points of the surface are given by the position of the corresponding point on the image. Using barycentric coordinates, we can easily improve the selection and the repositioning of points belonging to the surface to be textured.

ANIMATION OF THE BODY

A real-time virtual human is a virtual human able to act at the same speed as a real person. Virtual Reality, interactive television, and games require the bodies of real-time virtual humans. The generally accepted approach for body modelling is based on two layers: skeleton and skin; a third layer, cloth, could also be added. The skeleton layer consists of a tree-structured fixed-topology hierarchy of joints connecting limbs, each with minimum and maximum limits. Attached to the skeleton, there is the skin layer, which is responsible for generating the skin surfaces of the body (Thalmann, Shen and Chauvineau 1996). Animation of the virtual body is performed on the skeleton layer, and the two levels above are automatically computed by deforming or transforming vertices. This means that animation of the skeleton is not normally dependent on the other two layers and could be defined in very different ways.

Motion control is at the heart of computer animation. In the case of a virtual human, it essentially assists in describing, with respect to time, the changes to the joint angles of the hierarchical skeleton structure. Based on a general hierarchy manager library, a specialised library is built to model the human body hierarchy. This hierarchy is defined by a set of joints, which correspond to the main joints of a real human. Each joint is composed of a set of degrees of freedom, typically rotation and/or translation, which can vary between the values of two limits, based on real human mobility capabilities. Through a set of scaling methods applied to several points of this skeleton, it is possible to obtain a set of different bodies in terms of size (global scaling of body parameters) and in terms of characteristics (local scaling like spin, lateral or frontal scaling). The current virtual skeleton is a hierarchical model comprising a total of 32 joints corresponding to 74 Degrees of Freedom (DOF) including a general position–orientation joint.

Once the body is defined in terms of shapes and mobility, a global motion control system is used to animate the virtual human skeleton within 3D worlds. This is basically done by animating the joint angles over time. There are three distinct categories of skeleton motion control approach for virtual humans: motion capture in real time for driving an avatar, the use of predefined motions, and a computing dynamic model.

For the deformation of the body in real time, a trade-off is done between speed and realism. For applications imposing animation speed, typically, a polygonal representation is used: the skin that is wrapped around the skeleton is then represented with a fixed mesh divided at important joints where all the deformations are performed. Because no deformations are computed within a body part, i.e. between two joints, the virtual human appears 'rigid' and lacks realism. Moreover, visually distracting artefacts may arise at joints where two body parts are connected, e.g. when limbs are bent. There are some applications that stress visual accuracy. In such applications, the skin is generally computed from implicit primitives and a physical model is used to deform the body's envelope. Though this yields very satisfactory results in terms of realism, it is so computationally demanding that it is not really suitable for real-time applications. We use another approach, which combines some elements of the previous ones, allowing a good trade-off between realism and the rendering of speed. To exploit the high-end graphics capabilities, the body surface data is converted into triangle meshes. It is easy to construct a triangle strip from two adjacent cross-sections by simply connecting their points. Thus it is possible to construct an entire triangle mesh for each body part directly from the points of the contours. It is a bit more complicated to connect two different body parts, as the contours may have a different number of points, but the idea remains essentially the same. We eventually obtain a single, seamless body mesh.

The basic idea for fast deformations of human limbs and body torso is to manipulate the cross-sectional contours, thus transforming a complicated 3D operation into a 2D operation that is more intuitive and easier to control. A contour is by definition a set of points that lie in a plane. By setting the orientation and position of this plane, smooth deformations of the skin are achieved. Every joint in the skeleton is linked to a contour and it is ascertained that every joint lies in the plane of its contour when the skeleton is in the at-rest posture. The implementation is done using a real-time oriented 3D graphics tool kit called Performer that is available on all Silicon Graphics workstations.

ANIMATION OF THE HANDS

Hands have had very specific treatment in real-time human simulations. It is generally considered too expensive to use a deformable hand model, even considering the advantages it brings to the global visual result. This approach is closely linked to the optimisation approach generally used in virtual environments: the Level of Detail (LOD) (Funkhauser and Sequin 1993). In this approach, the importance of an object is mainly linked to its rendered size relative to the size of the final picture. According to this hypothesis, hands should not be given a lot of attention. We want first to justify the need for providing a real-time and accurate hand simulation model by briefly underlining the importance of hands in human simulation inside virtual environments. We will then show how we developed a dedicated simulation model for the hand, and then parametrised it in order to perform realistic hand simulation for real-time environments.

Hands in human simulation

Hands represent a very small part of the whole body. However their importance is not restricted to their size. Hand gestures can be classified into three main categories, which are described as follows (*Hand gestures for HCI*, 1996):

- **semiotic**: to communicate meaningful information and results from shared cultural experience.
- **ergotic**: associated with the notion of work and the capacity of humans to manipulate the physical world, to create artefacts.
- **epistemic**: allows humans to learn from the environment through tactile experience or haptic exploration.

These three categories show how important the hand is in simulating a realistic human when interacting with a virtual environment. The hand is both an effect and a sensor. It is a gateway between humans and their environment. This implies that hands are a centre of interest, and that, despite their size, many situations during the simulation will focus on them. In order to provide a convincing visual representation, they have to be appropriately modelled and simulated.

Hands concentrate a great number of Degrees of Freedom. Our skeleton model counts around 120 total DOF. Each hand contains 25 DOF, so both account for approximately 40 per cent of the total DOF. As a result, the hands are the most flexible part of the body, and the total number of possible postures and movements is very large. Thus, the hand may have a high level of deformation, concentrated in a very limited area. Moreover, a brief look at anyone's hands shows that, in addition to muscular action, skin deformation is controlled by the main hand lines associated with the joints of the skeleton.

Although the size aspect is important, we have shown that the importance of the hand, as part of a human simulation, requires more attention than it previously received, seen as a set of rigid articulated skin pieces. Although the hand is a part of the body, we have shown that its particularities, such as the importance of hand lines in relation to skin deformation, require a dedicated model. We propose a model for hand simulation, suited for real time, to be used in conjunction with the traditional rigid skin pieces approach.

The basic hand multilayer model

We have proposed a multilayer model for the simulation of human hands (Moccozet and Magnenat-Thalmann 1997), as a part of the HUMANOID environment (Boulic *et al*. 1995). The basic model, following the traditional multilayer model for an articulated deformable character, is subdivided into three structural layers (skeleton, muscle and skin). We combine the approaches of Chadwick, Hauman and Parent (1989) and Delingette, Watanabe and Suenaga (1993) to design the tri-layer structure deformation model for hand animation. The intermediate muscular layer, which maps joint angle variations of the basic skeleton layer on to geometric skin deformation operations, is based on a generalised Free-Form Deformations (FFD) (Sederberg and Parry 1986) model called Dirichlet Free-Form Deformations (DFFD). The structural muscle layer is modelled by a set of control points attached to the skeleton, which approximate the shape of the hand skin. The motion of the skeleton controls the displacement of the control points. Once the control point set is fitted to the current configuration of the skeleton, the generalised FFD function is applied to the triangle mesh that represents the geometric skin. We now briefly describe the main features of the basic geometric deformable model involved in the multilayer hand model, and how it is used to simulate muscular action on the skin. The main objective of the resulting model is to simulate how the muscles deform the skin and how hand lines control the skin deformations.

The relationship between the control points and the object to be deformed is based on a local coordinate system called Natural Neighbors (NN) or the Sibson coordinate system (Sibson 1980; Farin 1990; Sambridge, Braun and MacQueen 1995). The geometric deformation model is thoroughly described in Moccozet (1996). This local coordinate system makes it possible, for each vertex of the surface, to automatically define a subset of control points whose displacements will affect it. This subset is used to build a deformation function similar to the one defined for FFDs. The deformation function is defined inside the convex hull of the control points, and interpolates the displacements of the control points to the vertices of the deformed surface. Among all the properties of the resulting model it must be noted that there is no constraint on the location of the control points, nor on the shape of their convex hull. Moreover, there is no need to explicitly define the topology of the control points set. All of the FFD extensions can be applied to DFFD. Among them are two of particular interest to our hand simulation model. Weights can be assigned to control points and define Rational DFFD (RDFFD) (Kalra and Magnenat-Thalmann 1992) with an additional degree of freedom to control deformations. Direct surface manipulation (Hsu, Hugues and Kaufman 1992) can also be performed with the basic model without having to use any estimation method. As any location in the 3D space can be defined as a control point, assigning a control point to a vertex on the surface allows us to directly control its location: any displacement applied to the constraint control point is integrally transmitted to the associated vertex of the surface. Thanks to the flexibility introduced by our generalised FFD model, we provide an efficient tool to design an intermediate layer to interface the skeleton motion and the skin deformation, by simulating the behaviour of muscles and hand lines (see Figures 4.7–4.9).

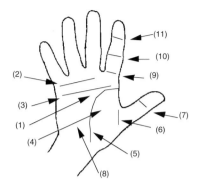

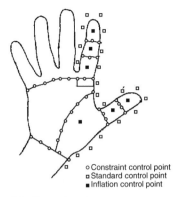

o Constraint control point
□ Standard control point
■ Inflation control point

Figure 4.7 Topography of the hand: (1) palm; (2) upper transversal line; (3) lower transversal line; (4) thenar eminence; (5) thenar line; (6) thumb first line; (7) thumb second line; (8) hypothenar eminence; (9) finger first line; (10) finger second line; (11) finger third line.

Figure 4.8 Control points arrangement.

(a) 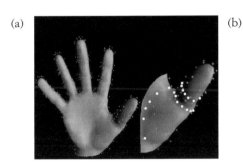 (b)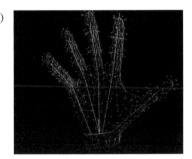

Figure 4.9 Control points set (a) and wrinkles design (b).

The set of control points is built in order to match a simplified surface hand topography. From the observation of a hand, especially its topography (Kapandji 1980), we derive the following basic data structure for our hand model, which we call a wrinkle, as shown in Figure 4.8:

- The wrinkle itself, which is a set of constraint control points. They are generally selected around the joint and form a closed 3D line. We call such points wrinkle control points.
- Two points among the wrinkle control points. They are used to define the axis on which the associated skeleton joint should lie. In this way, a skeleton model can be easily adapted to the hand's skin. This data allows an easy and realistic hand-skeleton mapping by defining an implicit skeleton to which the actual skeleton can be fitted.
- A mixed set of control points and constraint control points. They surround the upper part of the hand surface that will be affected by the rotation of the joint associated with the current wrinkle. We call these points influenced wrinkle control points, as they are influenced by the rotation of the wrinkle itself.
- One control point, called an inflation control point, which will be used to simulate inflation by the upper limb associated with the joint.

For each wrinkle, the muscle layer gets the joint angle variation from the skeleton layer. If the rotation angle is α, the wrinkle itself is rotated at an angle of $\alpha/2$, and the set of influenced control points is rotated at α. At the rest position, all control points have a weight of 1. When the joint angles vary, the weights of the inflation control points vary accordingly. Each point is placed on the mesh so that when its weight increases, it attracts the mesh, simulating the skin inflation due to muscle contraction.

Figure 4.7 shows the simplified surface hand topography we want to model and the important hand lines associated with the underlying skeleton joints. Figure 4.9 shows how control points, constraint control points and inflation control points are designed around the surface of the hand to build the control points set and the different wrinkles.

The optimisation approach

The complexity of our basic muscular deformation model is linked to:

- the degree of the deformation function; and
- the number of control points involved.

Our goal in this optimisation step is to provide a deformation function that is able to work at different levels, and introduce a Deformation Level of

Detail, in a way similar to Geometric Level of Detail. The optimisation is achieved by parametrising the deformation function with two features that constrain the function complexity. The real-time hand simulation model is working on a fixed Geometric LOD. This is motivated by the fact that a realistic hand shape requires a high resolution, and that performing deformations is only worthwhile for a minimum resolution of the deformed surface.

As for basic FFDs, the deformation function is a cubic function of the local coordinates. The properties of the deformation function make it possible to use it at lower degrees with a minimal loss in the continuity of the deformation. We can then choose between a linear, a quadratic or a cubic deformation function.

The total number of control points involved in the deformation of a vertex are not predefined and depend on the local configuration of the set of control points at the location of the vertex inside the convex hull of control points. The number of control points can be controlled and constrained between four control points up to the 'natural' number of NN control points. It must be noted that limiting the number of control points to four results in a continuity problem. This has to be considered as the price to pay for the gain in speed. Figure 4.10 shows, for the same hand posture, three different results with various constrained Sibson control points. Figure 4.10(a) shows an example of hand posture using the basic DFFD function, without any constraint on the number of Sibson control points. Figure 4.10(b) and (c) shows the same hand posture with the number of Sibson control points respectively constrained to 9 and 4. The figure of the hand contains around 1,500 vertices.

As a conclusion to the deformation function optimisation, we have shown that the complexity of the deformation function can be controlled and parametrised according to the degree of the function and the maximum number of control points used. If the deformation function name is DFFD, Deformation LODs can be defined by setting the two parameters Degree, and MaxControl of the DFFD function: DFFD (Degree, MaxControl). Degree can take its value between one, two and three, and MaxControl takes a value higher than four.

(a) Unconstrained   (b) 9 control points   (c) 4 control points

Figure 4.10 Hand posture without constraint (a), and with various constrained Sibson control points (b) and (c).

FACIAL ANIMATION

In our real-time human animation system, a face is considered as a separate entity from the rest of the body, due to its particular animation requirements. The face, unlike the body, is not based on a skeleton. We employ an approach that is different from body animation for deformation and animation of a face, based on pseudo-muscle design.

Developing a facial model requires a framework for describing geometric shapes and animation capabilities. Attributes such as surface colour and textures must also be taken into account. Static models are inadequate for our purposes; the model must allow for animation. The way facial geometry is modelled is motivated largely by its animation potential as considered in our system. Facial animation requires a deformation controller or a model for deforming the facial geometry. In addition, a high-level specification of facial motion is used for controlling the movements.

The facial deformation model

In our facial model the skin surface of a human face, being an irregular structure, is considered as a polygonal mesh. Muscular activity is simulated using Rational Free-Form Deformations (RFFD). To simulate the effects of muscle actions on the skin of a virtual human face, we define regions on the mesh corresponding to the anatomical descriptions of the regions where a muscle is desired. For example, regions are defined for eyebrows, cheeks, mouth, jaw, eyes, etc. A control lattice is then defined on the region of interest. Muscle actions to stretch, expand, and compress the inside geometry of the face are simulated by displacing or changing the weight of the control points. The region inside the control lattice deforms like a flexible volume according to the displacement and weight of each control point. A stiffness factor is specified for each point, which controls the amount of deformation that can be applied to the point; a high stiffness factor means that less deformation is permissible. This deformation model for simulating muscle is simple and easy to perform, natural and intuitive to apply and efficient to use for real-time applications.

Facial motion control

Specification and animation of facial animation muscle actions may be a tedious task. There is a definite need for higher level specification that would avoid setting up the parameters involved for muscular actions when producing an animation sequence. The Facial Action Coding System (FACS) (Ekman and Friesen 1978) has been used extensively to provide a higher level specification when generating facial expressions, particularly in a non-verbal communication context. In our multilevel approach (Figure 4.11),

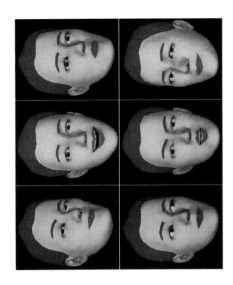

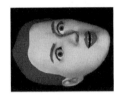

Figure 4.11 Different levels of facial motion control.

we define basic motion parameters as Minimum Perceptible Actions (MPAs). Each MPA has a corresponding set of visible features such as movement of eyebrows, jaw, or mouth and others occurring as a result of muscle contractions. The MPAs are used for defining both facial expressions and visemes.[2] There are 65 MPAs, such as open-mouth, close-upper-eyelids or raise-corner-lip, used in our system. This makes it possible to construct practically any expression and viseme. At the highest level, animation is controlled by a script containing speech and emotions along with their duration. Depending on the type of application and input, different levels of animation control can be utilised.

There are three different input methods used for our real-time facial animation module: video, audio or speech, and predefined actions.

Video Input

This requires facial feature extraction and tracking from the image sequences of the video input. We use an improved method developed by Magnenat-Thalmann, Kalra and Pandzic (1995) that returns an array of MPAs corresponding to the extracted facial feature movements. Mapping of the parameters of the extracted features is obtained as the displacement vectors are based on some ad hoc rules. The extraction method relies on a 'soft mask', which is a set of points adjusted interactively by the user. Recognition and tracking of the facial features is based on colour-sample identification, edge detection and other image-processing operations. The capture and tracking rate of features is about 20 frames/sec. on an SGI O2 workstation.

Audio/Speech Input

We rely on an external program (*ABBOT (Demo), Speech Recognition System*) for segmentation of the audio into phonemes, with the duration of each phoneme. In the absence of audio input we use text as input and obtain the phonemes by using a text to phoneme module developed by the University of Edinburgh (*Festival Speech Synthesis System*). Each phoneme is translated into a viseme, which is decomposed into several MPAs. Visemes are defined as a set of MPAs that are independent of the facial model and can be applied to any face.

Predefined Actions

Real-time animation of a face can also be performed using a series of predefined expressions and visemes. Here, the specification is at a higher level – an *action*, that has intensity, duration and start time. An action may be an emotion (surprise, anger, etc.), head gestures (nodding, turning, etc.), and sentences (combinations of words defined with phonemes). The actions

72

are decomposed into an array of MPAs and deformation is performed accordingly for each frame during the animation.

Synchronisation

In order to have synchronised output of the MPA arrays from different sources (e.g. emotions from video and phonemes from audio-speech) at a predefined frame rate (Fd, generally 25 frames/sec.) with the acoustic speech, a buffer or stack is introduced for each source of MPAs. An initial delay is caused if the frame rate of one source is less than Fd (see Figure 4.12).

It is assumed that for each MPA source the frame rate is known (e.g. F1, F2). The intermediate frames are added using interpolation/extrapolation of the existing computed frames in each buffer to match the frame rate Fd. The MPA array from each buffer goes to the composer, which produces a single stream of MPAs for the deformation controller where a face is deformed. The deformation process for each frame on average takes less than one-fortieth of a second on a fully textured face with about 2,500 polygons on an SGI O2 workstation.

Composition

As the animation may involve simultaneous application of the same MPA coming from different types of actions and sources, a mechanism to compose the MPAs is provided. A weight function is defined for each MPA in an action. It is a sinusoidal function with a duration relating to an action, generally considered as 10 per cent of the total duration of the action. This

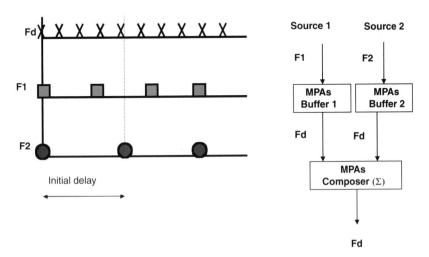

Figure 4.12 Synchronisation of MPA streams.

provides a smooth transition with no jump effect, when there is overlap of actions with the same MPA.

THE ANIMATION FRAMEWORK

One unique feature of our real-time simulation of virtual humans is the close link between modelling and animation. Here, modelling does not mean just constructing geometrically passive objects, but includes structure and animation capabilities. In our system the modelling is driven by the animation potential of the body parts. In addition, the modelling facilitates easy control of multiple levels of detail. This point is a big asset for real-time applications, particularly when many virtual humans inhabit the same virtual world. By real-time applications, we mean to be able to display at least ten frames per second while using the program.

As previously stated, the system can be broadly separated into three units: modelling, deformation and motion control as shown in Figure 4.13.

Modelling provides necessary geometrical models for the body, hands and face. The *BodyBuilder* program is used for modelling the body surface. These surface contours are associated with the skeleton segments and joints. The skeleton is used as a support in the generation of motion. For hand creation, a default template hand can be used. Designers can then modify this template to build a specific hand. The default hands are associated with the skeleton to provide postures used for real-time animation. Both local and global transformations are performed in the *Sculptor* program as described above. Similarly, the face generally starts from a generic model. This model includes structure information provided by the definition of regions. Modifications of the face, however, are done in a manner that retains the structure. A method is developed where automatic face reconstruction is done using two orthogonal views of pictures.

Deformations are performed separately on different entities (body, hands, face). They are based on the model used for each part. The choice of a different model for each entity is motivated by the particular animation requirements of each entity in real time, which have been elaborated in earlier sections. Different entities are assembled into a single skin envelope using the *DODY* (*Deformable bODY*) library. Handling and managing the deformations of each entity are also performed in the *DODY* library.

Motion control is used for generating and controlling the movements of different entities. For motion control, the body and face are separated. The hands are included in the body as they are also skeleton-based. The skeleton motion can be generated using an interactive software program developed at LIG-EPFL, called *TRACK*, in which some predefined actions can also be combined. A motion capture module is also available for real-time motion capture of the body. Similarly, facial movements can be generated in terms

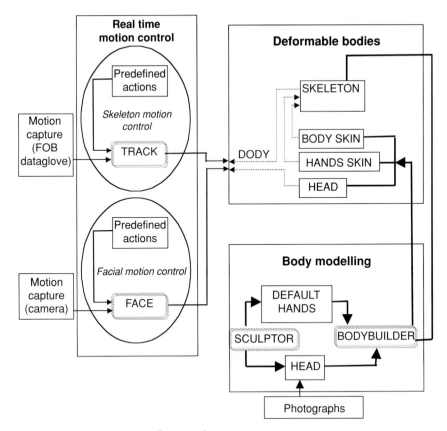

Figure 4.13 The animation framework.

of expressions/phonemes in an interactive *FACE* program. Direct motion capture from a real face is also possible. The body motion, in terms of angular values of joints, and face motion, in terms of MPAs, are passed to the *DODY* library to perform the appropriate deformations for animation.

A higher level motion library is also provided in which motion is designed as a set of actions for the different entities. These actions can be blended and simultaneously applied. This offers a high-level programming environment suitable for real-time application.

CASE STUDY: CYBERDANCE

CyberDance is a new kind of live performance providing interaction between real professional dancers on stage and virtual ones in a computer-generated world. It can, therefore, be thought of as a form of HyperReality. This performance uses our latest development in virtual human representation (real-time deformation) together with the latest equipment in Virtual Reality (for motion tracking).

Our first performance was made for the Computer Animation film festival in September 1997, in Geneva. It was an 18-minute live show with eight professional dancers and giant screens for computer-generated images. The show was divided into three parts and represented the creation of the 'second world'. Later on, we performed at the Creativity Institute at Zermatt, and at another site in Geneva.

In the first part, the virtual world recreated the planet Earth in the universe and the choreography reflected the evolution of different styles of music through time. In the second part, virtual humans were added to the virtual world and one real dancer was tracked to animate his virtual clone in real time, represented by a fantasy robot. Figure 4.14 shows snapshots of the live performance using motion capture. We can see the dancer tracked on stage, while the result of this tracking was used to construct the virtual robot and displayed in real time. The audience was able to see both the real and the virtual dancer at the same time. Then, in the last part, the virtual actors danced following a given choreography. In Figure 4.15, we can see the three virtual dancers following the choreography recorded using a motion capture device. The same choreography was used for the three clones sharing the same environment.

CONCLUSION

We have shown in this chapter how to model and animate believable and realistic virtual humans in virtual environments. These virtual humans are real-time, they may assume any position, and may be used for interactive TV applications, simulation, and shared Virtual Environments and, of course, in HyperReality. Different techniques are employed to suit the requirements of different body parts. An animation framework integrates different modules for modelling, deformation and motion control. A case study, CyberDance, was presented to demonstrate the potential of such a system in HyperReality. Further research includes the elaboration of a user-interface for real-time simulation and improvement of the visual quality of the simulated individuals. Increasing realism requires revising and improving our methods, although the results should not be qualitatively very different. We are working on the real-time simulation of hair and deformable clothing, and on a variety of autonomous behaviours. We are also developing methods for accelerating the cloning process.

Figure 4.14 Motion capture for real-time dancing animation of a virtual clone.

Figure 4.15 Professional dancers and three virtual actors dancing at the same time.

NOTES

1 Modules for body (excluding hands and face) deformation and animation have been developed primarily at LIG-EPFL; we share the testing and integration of all the modules.
2 Visemes are defined as the animated shape of the face resulting from the motion of the mouth (lips and jaw) and corresponding to one or more phonemes (different phonemes may correspond to the same viseme).

ACKNOWLEDGEMENT

With thanks to Jean Claude Moussaly, Marlene Poizat and Nabil Sidi Yacoub from MIRALab, who designed the virtual humans and the CyberDance environment. We thank Prof. Daniel Thalmann, Director LIG-EPFL and his research team for their collaboration.
This research is funded by SPP Swiss research program, the Swiss National Science Foundation and eRENA European project for the CyberDance project.

REFERENCES

ABBOT (Demo), Speech Recognition System, Cambridge, UK. URL: http://svrwww.eng.cam.ac.uk/~ajr/abbot.html

Boulic, R., Capin, T., Huang, Z. *et al.* (1995) 'The HUMANOID environment for interactive animation of multiple deformable human characters', *Proceedings Eurographics '95, Maastricht*: 337–348, Oxford: Blackwell.

Chadwick, J.E., Hauman, D. and Parent, R.E. (1989) 'Layered construction for deformable animated character', *Proceedings Siggraph '89*: 243–252, New York: ACM Press.

Delingette, H., Watanabe, Y. and Suenaga, Y. (1993) 'Simplex based animation', in N. Magnenat-Thalmann and D. Thalmann (eds), *Proceedings Computer Animation '93*: 13–28, Heidelberg: Springer Verlag.

Ekman, P. and Friesen, W.V. (1978) *Manual for the Facial Action Coding System*, Palo Alto, CA: Consulting Psychology Press.

Farin, G. (1990) 'Surface over Dirichlet tessellations', *Proceedings of Computer Aided Geometric Design*, 7: 281–292, Amsterdam: North-Holland.

Festival Speech Synthesis System, University of Edinburgh, UK. URL: http://www.cstr.ed.ac.uk/projects/festival.html

Funkhauser, T.A. and Sequin, C.H. (1993) 'Adaptative display algorithm for interactive frame rates during visualization of complex virtual environments', *Proceedings Siggraph '93*: 247–254, New York: ACM Press.

Hand gestures for HCI, Hand Centered Studies of Human Movement Project, Technical Report (http://fas.sfu.ca/cs/people/ResearchStaff/amulder/personal/vmi/HCI-gestures.htm), School of Kinesiology, Simon Fraser University, 1996.

Hsu, W., Hugues, J.F. and Kaufman, H. (1992) 'Direct manipulation of free-form deformations', *Proceedings Siggraph '92*: 177–184, New York: ACM Press.

Information International Inc. (1982) SIGGRAPH film festival video review, New York: ACM Press.

Kalra, P. and Magnenat-Thalmann, N. (1992) 'Simulation of facial muscle actions based on rational free-form deformations', *Computer Graphics Forum*, 2, 3: 65–69.

Kapandji, I.A. (1980) *Physiologie articulaire, Tome 1*: 45–58, Maloine SA Editions.

Lee, W., Kalra, P. and Magnenat-Thalmann, N. (1997) 'Model based face reconstruction for animation', *Proceedings of MultiMedia Modelling '98*: 323–338.

Litwinowicz, P. and Miller, G. (1994) 'Efficient techniques for interactive texture placement', *Proceedings Siggraph '94*: 119–122, New York: ACM Press.

Magnenat-Thalmann, N., Bergeron, P. and Thalmann, D. (1982) 'DREAMFLIGHT: A fictional film produced by 3D computer animation', *Proceedings of Computer Graphics, Online Conference*: 353–368.

Magnenat-Thalmann, N. and Kalra, P. (1995) 'The simulation of a virtual TV presenter', *Proceedings of Pacific Graphics '95, World Scientific*: 9–21.

Magnenat-Thalmann, N., Kalra, P. and Pandzic, I.S. (1995) 'Direct face-to-face communication between real and virtual humans', *International Journal of Information Technology*, 1, 2: 145–157.

Magnenat-Thalmann, N. and Thalmann, D. (1995) 'Digital actors for interactive television', *Proceedings of IEEE: Special issue on digital television, Part 2, July*: 1,022–1,031, Los Alamitos, CA: IEEE Computer Society Press.

Moccozet, L. (1996) *Hands modelling and animation for virtual humans*. Thesis report, MIRALab, University of Geneva.

Moccozet, L. and Magnenat-Thalmann, N. (1997) 'Dirichlet free-form deformations and their application to hand simulation', *Proceedings of Computer Animation '97*: 93–102, Los Alamitos, CA: IEEE Computer Society Press.

Sambridge, M., Braun, J. and MacQueen, H. (1995) 'Geophysical parametrization and interpolation of irregular data using natural neighbors', *Geophysical Journal International*, 122: 837–857.

Sannier, G. and Magnenat-Thalmann, N. (1997) 'A user-friendly texture-fitting methodology for virtual humans', *Proceedings of Computer Graphics International '97*: 167–176, Los Alamitos, CA: IEEE Computer Society Press.

Sederberg, T.W. and Parry, S.R. (1986) 'Free-form deformation of solid geometric models', *Proceedings Siggraph '86*: 151–160, New York: ACM Press.

Shen, J. and Thalmann, D. (1995) 'Interactive shape design using metaballs and splines', *Proceedings of Implicit Surfaces 1995*, Grenoble, France: Eurographics.

Sibson, R. (1980) 'A vector identity for the Dirichlet tessellation', *Mathematical Proceedings Cambridge Philosophical Society '87*: 151–155.

Slater, M. and Usoh, M. (1994) 'Body centred interaction in immersive virtual environments', in N. Magnenat-Thalmann and D. Thalmann (eds), *Artificial Life and Virtual Reality*, Chichester: Wiley.

Thalmann, D., Shen, J. and Chauvineau, E. (1996) 'Fast human body deformations for animation and VR applications', *Proceedings of Computer Graphics International '96*: 166–174, Los Alamitos, CA: IEEE Computer Society Press.

5

ARTIFICIAL LIFE IN HYPERREALITY

Katsunori Shimohara

Editors' introduction
Besides being where the real and the virtual interact, HyperReality is where human
and artificial intelligence interact. This chapter looks at the implications of this
from the perspective of research on evolutionary systems for brain communication,
based on principles of communication inspired by the mechanisms of genetics and
evolution as they occur in nature. Seeking seamless interaction is more than making
the mechanics of interaction between real and virtual components of coaction fields
appear natural, as when a virtual person shakes hands with a real person or a
real bat hits a virtual ball. It also means that communication between the virtual
and the real must be as fluent and dynamic as it is in the single dimension of phys-
ical reality. How does one develop communication that appears seamless when the
interaction is between human and artificial intelligences? Katsunori Shimohara
examines the possibility of artificial pets in HyperReality as a way of exploring
human-to-computer communication. He sees the development of such communication
as being dependent on artificial intelligence acquiring autonomy, creativity and feel-
ings and giving us then artificial life.

   Shimohara studies artificial life in terms of both hardware and software, and
sees the development of artificial life as evolutionary, subject to the principles of
natural selection and lying within an interactive hierarchy of artificial life forms.
Real life originates and evolves in real reality but it can enter into virtual reality
via the portals of coaction fields. Artificial life can be created and evolve in virtual
reality where it can be embodied in avatars. Shimohara describes how artificial life
can also enter into physical reality in robotic form. Will it also evolve there?

INTRODUCTION

One of the benefits of HR should be to enrich human communications by
extending their coaction fields to virtual worlds with diverse chances and
environments where they can interact with artificial life in virtual space or
real space.

One idea to involve people naturally and to motivate spontaneous participation in HR is to use a virtual character and/or creature as a pet and to make it play the role of attracting people to interact with itself in the HR world. Such a character and/or creature need not be restricted to a virtual world. The virtual character or pet can appear to a user as a physical robot in the real world. It will not be difficult to realise a mechanism that enables the virtual character (or pet) to traverse between a virtual world and the real world.

When we focus on communications between people and virtual characters, two research issues come up: what people desire in communications with virtual creatures, and how to implement attractive virtual creatures.

Even though the first issue is significant, it is a difficult one to discuss in general. However, by relating it to the second issue, the following example suggests a way forward. A child playing with a toy tends to become quickly bored, while people with a pet dog or cat are seldom bored in their interactions with their pet. Before long, pre-programmed toys become boring and lose their interest, whereas living, or apparently living, creatures can continue to hold their attractions for people.

In other words, we can postulate that it is vital to enable improvisational interactions between people and virtual artificial characters, and for that purpose virtual artificial creatures should be able to generate information autonomously. That is, they must be rich in autonomy and creativity. Such characteristics could emerge in an environment such as HyperReality, which seems to be related to evolutionary systems based on artificial life.

In this chapter, envisaging intelligent virtual characters and their robotic extensions in the near future, we consider how they can activate communicative interactions with people. Second, the concepts of artificial life and research topics concerning evolutionary systems are introduced as promising methodologies to achieving autonomy and creativity in virtual characters and robots.

VIRTUAL CHARACTERS/ROBOTS IN THE NEAR FUTURE

Consider having a virtual character/robot for a pet, a virtual character/robot that would communicate with us as a partner to back us up. I want to take the following pages to think imaginatively about the shapes that virtual characters/robots will take in the near future.

Virtual characters/robots as pets

When we refer to 'virtual pets' or 'pet robots', we are imagining virtual characters/robots that can be raised by their users through continued interaction, or characters/robots that, through interaction, mature, develop, and evolve alongside their users.

The great popularity of the electronic pet, the tamagotchi, points to the fact that people have an untiring interest in these so-called nurturing games. But even without the example of the tamagotchi, it is easy to see in both pet-raising and bonsai-cultivating a basic human need to interact with other living things. In one sense, we can understand this need as a manifestation of our urge for existential expression; in some form or other we want to leave behind, in the external world, an expression of our existence through the interactions and influences we have had on other living beings. Fundamentally, humans are creatures that seek interaction with others, creatures that desire communication.

Relying on this human desire and predisposition, a virtual character/robot needs to be able to behave independently, just as a living pet does, if it is to gain a human being's willing involvement. An evolutionary methodology is by far the most effective means of realising this end. With an evolutionary methodology, one can automatically generate behaviour that is generally in step with the system but can occasionally be quite unusual. It is this unexpected, unusual behaviour that, in a sense, stimulates the user's imagination, potentially increasing the user's desire to affect the outside world.

Because humans can relate at a corporeal level to a robot that they can touch directly, these robots are qualitatively different from other robots or electronic pets that exist solely within electronic space or as software agents. Within the quick-paced cyberisation that is under way today, the existence of robots able to share bodily sensations and corporeality will only increase in importance.

In addition, humans and virtual characters/robots in their ultimate form could share the sense of being in the same body. We can easily imagine the development of pet-form characters/robots as wearable devices whereby humans and their characters/robots are in constant direct contact. Humans would naturally communicate their physical and mental states to the character/robot. The character/robot would not only receive the human's conscious, intended directions and commands, but, in using information about the user's physical and mental state, it could adapt itself smoothly to its user. With the robot's connection to networks, it would be easy to develop a system in which the robot detects irregularities in the user and the virtual character automatically transmits emergency messages.

Communication between humans and virtual characters/robots

The computers we come into contact with today respond to our commands. It is the same with the current generation of virtual characters/robots.

Fundamentally, they move in compliance to our commands and only after they have received our commands. If it is complicated work, they begin work only after they have finished the negotiations required to undertake it. That, however, is not how humans work. Human beings often begin an activity, then stop to think about its meaning, and then subsequently restart the activity and begin to communicate. A good example is when we greet someone with a bow of the head. A bow happens when the two parties tip their heads forward at almost exactly the same time.

In such a way, by acting first and assigning meaning retroactively, we can reduce the costs of communication. For example, it's often the case that when we are in the passenger seat of a car giving directions, without thinking we simply say, 'That way!' In saying this a speaker is assuming that the listener has comprehended the speaker's point of view and context. If the listener goes in the wrong direction, then we will carefully set him or her straight, but if the listener goes in the right direction, we have communicated the information at little cost. This mode stands in opposition to communication with today's robots, whose commands must be fully elaborated in advance.

For virtual characters/robots to enter human society and serve as supports in our daily lives, it is imperative that they be able to assign meaning to activity retroactively. A virtual character/robot that needed to negotiate each and every detail with us, far from being a support, would be a hindrance in our daily lives. The technology that is required in virtual characters/robots is one where humans and virtual characters/robots can put their bodies to practical use, assigning meaning to each other's actions. This step is an evolution to the next stage where virtual characters/robots can cohabit with us as partners.

To make possible the kind of rich communication between humans and robots described above, there must be more than machine-like interaction; there must also be interaction based on emotions. If we define emotions as values and evaluative structures that emerge out of relations with others, then our technological challenge is to equip virtual characters/robots with mechanisms for artificial emotions; devices that will foster this capability as a sort of ruse. What becomes crucial here is whether the robot can distinguish itself from others; at a more complex level, the border between self and other needs to be made flexible. We humans freely adjust this flexible border between self and other, making choices and decisions about our actions according to the conditions and times in which we find ourselves. For example, we locate ourselves as part of a family, group, organisation, collective, or finally of a nation or race that we belong to. Thus locating ourselves, we consider issues, communicate ideas, and act upon the world. Or, in other cases, we can each produce an objectified self within our own individuality that we use to reflect upon our own actions.

This kind of mechanism is necessary for the virtual character/robot to foster an identity by itself. If virtual characters/robots did have this kind of artificial emotional mechanism and sense of self and others, they would be able to enrich and diversify communication with people, just as we humans do with each other.

Virtual characters/robots as communication partners

One inherent characteristic of a virtual character/robot is that a robot in the physical world is connected to the networked world. With robots connected to networks, there will be a dramatic expansion of a new world of communication between humans and robots. These kinds of robots will be able to traverse freely between the real world and cyberspace while existing simultaneously in both worlds, at times as physical entities, at times as software-like network agents. Each a version of the other, they can be substituted interchangeably.

Their physical bodies (their hardware) are something akin to open vessels. On the road you can call up your own virtual character's mind (its software) by borrowing another vessel to connect to it over a network. With networking, the virtual character/robot can be brought along wherever a person is going, just as a virtual character/robot can be called up from anywhere.

Conversely, if these open vessels (bodies) were to exist like public telephones now do in every part of the world, then humans would be able to use networks to move instantaneously across the world through their robots. By borrowing the sensory perceptions of a robot and its body, one could visit a street corner somewhere and mix with the local inhabitants. Over networks, virtual characters/robots could assist in expanding a human's own bodily experiences; in other words, there could be tele-existence virtual characters/robots.

The evolution of virtual characters/robots

Pet-form communication virtual characters/robots, combining aspects of a pet with those of an information communications device, present one of the most promising possibilities for making information communications environments for the twenty-first century that anyone can use. They promise to enrich communication between people and to open up new possibilities for communication between virtual characters/robots and humans.

There are diverse developments possible in pet-form communication virtual characters/robots, and we can expect that they will provide new business opportunities. The painstakingly nurtured mind of a virtual character/robot, in other words the software that generates its adaptive behaviour, can be downloaded over networks.

What kind of virtual character/robot it will mature into, however, is dependent on its relations with its users. Thus, there may well be pet-form communication virtual character/robot training centres to retrain those virtual characters/robots damaged during their upbringing, and pet character/robot stores may even emerge. Good characteristics that have emerged through nurturing and training by many individuals can be passed down from generation to generation, and two pet characters/robots with superior characteristics can be mated, leading perhaps to the development of a pet character/robot lineage. What we will see here is not simply deployment of an evolutionary methodology as a means to grow, develop, and evolve adaptive behaviour; it is rather a question of making the pet characters/robots themselves evolve.

The character/robot's evolution cannot be limited to software. When interfaces have been standardised, we will be able to choose our CPU, choose our sensors, choose the actuator (servo), and assemble our robot just as these days we can assemble a build-it-yourself computer. And the day is not so far off when human hands will reproduce virtual characters/robots; with the unification of different standards and a plentiful supply of parts, character/robot blueprints will soon be on web pages. With the choice of any two such web pages, a new character/robot will be born. And when humans add new improvements in characters/robots, they will be contributing to the evolutionary process of the shape of characters/robots. If in their process of reproduction characters/robots become capable of using certain devices without the intervention of human beings, we might well see virtual characters/robots that are living within a human environment actually evolve.

In closing this section, I have attempted to stimulate our imagination as we think about the shape of virtual characters/robots in the near future and to explore possibilities for new kinds of communication that virtual characters/robots and humans are creating together. This sort of entity will play an important role in HyperReality.

EVOLUTIONARY SYSTEMS AS ARTIFICIAL LIFE-RELATED TECHNOLOGIES

At the Evolutionary Systems Department, ATR Human Information Processing Research Laboratories in Kyoto, Japan, we are researching the possibilities of building 'computers that can generate information', that is, information processing systems that are rich in autonomy and creativity.

The ultimate information processing system is, of course, the human brain, which recognises and understands the environment through information received from its sensors and generates information in the form of actions and behaviours, which in turn change the environment.

In our research, we apply the ideas and methodologies of artificial life (ALife) to the modelling of brain functions. The following sections describe the research concepts we use in working toward our goal of creating an artificial brain as an evolutionary system that can not only develop new functionality spontaneously but can also grow and evolve its own structure autonomously (Shimohara 1994).

An artificial brain as an evolutionary system

One of the main steps towards the goal of achieving an ideal interface between people and computers is to achieve a level of human-to-computer communication similar to that of human-to-human communication, or, more precisely, brain-to-brain communication. To this end, it is important to use computer science and engineering techniques to model certain functions of the brain, such as the ability to learn, memory, flexibility, adaptability, autonomy and creativity. Since we are aiming to create 'computers that can generate information', we will focus on autonomy and creativity.

Many studies of brain function modelling have been carried out with a variety of techniques and from the perspectives of various disciplines. For example, remarkable progress has been made recently in neurophysiology and cerebral cortex research in pursuit of an understanding of the functions of nervous tissues and their anatomical connections. However, it will be a long time before the brain is well understood if brain researchers limit themselves to traditional analytical techniques (i.e. to understand the whole, first understand the parts). On the other hand, in the synthetic field of artificial neural networks, which uses computational models based on biological nervous systems, many studies have focused their attention on the ability of such networks to learn and optimise under constraints. However, even with neural networks, it is usually human designers who determine the structures of the artificial neural nets, who prepare training data for them, and who supervise and control their training processes and behaviours. Therefore, it is fair to say that the field of neural networks is still far from creating autonomy and creativity.

One basic idea of the ALife-based approach is not synthesis by analysis but rather analysis by synthesis, as will be described later. The synthetic approach is one of the main characteristics of ALife and can thus be regarded as complementary to traditional analytical and anatomical approaches.

Before explaining the ideas and methodologies of ALife, I would like to introduce the new information processing systems we are aiming to create that will be rich in autonomy and creativity. Such information processing systems are called artificial brains and are examples of evolutionary systems. By evolutionary system, we mean that the system will be able not only to spontaneously develop new software but also to autonomously grow and evolve its own hardware structure.

What will an artificial brain as an evolutionary system be like? We do not intend to merely mimic a biological brain in its function and structure but wish to create an artificial brain that will be superior to the biological brain in certain respects.

In the mammalian brain, enormous quantities of neurons are produced both before and after birth. Many of them die, but still many billions remain. In the infant brain, the remaining neurons grow by spreading and extending their dendrites and axons, gradually forming networks of neurons by connecting axons and dendrites through synapses. From then on, neurons continue to die until the death of the individual. Human infants are very flexible and creative in their thinking because the structure of their growing brains is correspondingly flexible. The plasticity of synapses allows them to be influenced by experience and to learn by modifying themselves little by little. However, the brain ultimately dies when the body dies. Parents can only leave their genes to their offspring, no matter how hard they study or how much they experience, although of course they can teach and write books. The evolution of the biological brain has followed this repetitive pattern of production, growth, modification and death for countless generations.

However, if an artificial brain can be built, it should be possible to make part or all of it return to an infant brain state whenever flexibility is needed. This means it could restructure its neural networks and increase the number of its neurons whenever necessary. While a biological brain is limited to a given size so that a body can support it, an artificial brain need not be so limited. Also, an artificial brain would not need to die! An artificial brain could evolve a new part, leaving the results of learning and experience intact, and could add a new part to itself step-wise. For example, a reptilian brain part could be added to a fish brain, and a mammalian brain part could be added to the reptilian brain part, and so on. Also, it might be possible to build artificial brains that self-replicate and can evolve separately, thus forming a society of artificial brains, which could be connected to each other through the information superhighway.

A paradigm shift to life-like and society-like information processing

Artificial life (ALife) is a new research paradigm in which the essence of life is investigated from an informational and computational viewpoint. What is life? This issue could be debated eternally, but we can recognise the characteristic functions and behaviours of life and living systems when we see them in nature. ALife is a challenging new field that investigates the concept and logic of life as it could be, as well as providing a better understanding of life as we know it, by synthesising life-like phenomena in artificial media using informational and computational techniques (Langton et al. 1992).

From an engineering viewpoint, ALife can be regarded as a possible framework for achieving artificial systems that possess the advantages of living systems, such as autonomy, adaptation, evolution, self-replication, and self-repair. A typical example is the epoch-making work of Tom Ray who showed that computer programs can evolve in the virtual world of a computer (Ray 1992). This encourages us to explore the possibility that a computer itself might evolve.

One key concept of ALife is emergence. Emergence is explained as a system behaviour or a process in which a global order or state is spontaneously generated through local interactions of lower-level components obeying certain rules. Another key concept of ALife is the notion that the global order or state influences the behaviours and interactions of the lower-level components and makes them change. Chris Langton, the father of ALife, advocates collectionism, which combines these two key concepts. Collectionism is thus a conceptual framework that models a combined bottom-up emergence, a top-down influence, and how they functionally interact with each other, as shown in Figure 5.1.

When considering autonomy and creativity in an information system, the ALife methodology can be explained as an information processing mechanism in the following way.

First, prepare a collection of elements and a framework in which they are to interact with each other. Some of these elements are activated by a stimulus from the environment, which triggers their interaction. Through such interactions, a kind of whole (whether in terms of organisation, structure, order, network, or global state) spontaneously emerges. This whole can change due to its activation of other elements. Also, it is important to provide some mechanism whereby elements themselves can make changes.

We intend to use the above-mentioned information processing mechanisms in our research. We have adopted two approaches: (a) life-like modelling and (b) social modelling, in addition to the conventional modelling of nervous systems, e.g. neural networks. In life-like modelling, the system should have a self-assembling (embryological) capability, similar to living systems in nature, in order to allow its structure and components to change.

In social modelling, the system should have a micro-macro dynamics whereby a global or macroscopic order/state emerges through the behaviour and interaction of microscopic components. In turn, the behaviour of the microscopic components is influenced by the macroscopic state. Thus, two directions of interaction – bottom-up, from micro to macro, and top-down, from macro to micro – make it possible for systems to change interdependently.

An individual body is a complex system composed of many organs and an enormous number of cells. The cells combine to form an organ, and many organs combine to form a harmoniously functioning body. When we observe the behaviour of an individual cell or organ, it appears as an informational element that behaves with a high degree of autonomy. On

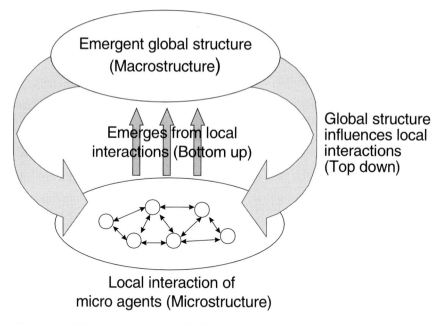

Figure 5.1 Collectionism: a basic ALife concept.

the other hand, when we observe cells or organs at a higher level, they function within a harmonious and integrated system, which is based on the interactions of informational elements such as nervous, metabolic and immune networks. Viewing a system as a harmonious aggregate of components can be extended to insect swarms or to human societies that consist of many individuals. It is important to note here that most of the elements at every level are not immutable but can change their number and function.

In summary, based on the above discussion, we propose here life-like and society-like information processing schemes in order to build an information processing system that is rich in autonomy and creativity. This proposal involves certain shifts in design principles, e.g.:

- From centralised to DECENTRALISED
  The behaviour of the system will not be centrally controlled but will emerge from the interaction of local components.
- From optimised to COLLECTIVE and REDUNDANT
  Functions should be generated and realised as situated and self-organised collective behaviours of the system.
- From fixed to FLUID
  Processing elements should emerge/disappear, replicate/die, combine/

separate, and then autonomously change through such mechanisms as metabolism and/or natural selection.

As fundamental research topics concerning evolutionary systems, software and hardware evolutions are described in the following sections.

Software evolution: evolving programs

This section introduces research on an evolutionary model of software evolution in which computer programs autonomously develop new functionality by utilising errors, which may occur spontaneously and/or interdependently, as changes for rewriting their own programs.

Tom Ray's 'Tierra' was an epoch-making work showing that programs can evolve in the virtual world of the computer (Ray 1992). The significance of this work was not only in raising and stimulating interest in ALife, but also in reaching a new frontier of information processing. That is, it proved the possibility of software evolution, where programs that are supposed only to perform their programmed instructions can autonomously rewrite themselves so that they work as new programs.

Tierra, which means earth in Spanish, is a kind of virtual world created in a computer in order to observe functions of self-replication and open-ended evolution. The aim is not to mimic life but to synthesise life. This is an attempt to create evolutionary processes, where digital organisms diversify and increase in complexity, in a computer. In other words, this project attempts to use evolution by natural selection in the medium of the digital computer to generate complex and intelligent software.

Living organisms in nature use energy provided by the sun, produce materials using the energy, and compete for resources such as light, food, space, and so forth. Natural selection operates through differences among individuals in their rates of survival and replication. In this way, evolution causes changes in the genetic information in the population of replicators. Genetic information is a series of instructions that describe its organism and can be regarded as a self-replication algorithm. Based on an analogy to the above, therefore, we can postulate a digital organism in which genetic information is just a self-replicating program composed of a series of machine codes. Accordingly, a virtual world is created in a computer, where digital organisms can self-replicate and evolve using CPU time as energy and its memory space as the environment for self-replication. In the sense that an organism itself in Tierra consists of genetic information and does not form any phenotype, Tierra is regarded as resembling the world of self-replicating ribonucleic acids (RNAs).

In this virtual world that models mutation and natural selection, self-replication starts with a single ancestor seed organism and is endlessly repeated. Mutation makes changes in their genetic information, i.e.

programs, and enables the formation of new species, which have different genetic information. Thus, natural selection through competition for energy and space allows species to evolve, and then an ecological system composed of diverse species with different genomes emerges spontaneously.

An animation image of an ancestor (a hand-written self-replicating program) is shown in Figure 5.2(b). An ancestor has a processor represented as a candle, corresponding to a kind of lifetime in the sense that it works as an energy resource for a limited period. The code consists of three genes – self-examination for measuring its size in memory, a reproduction loop for allocating a block of memory of the measured size for its replication, and a copy procedure with which the entire genome is copied into new memory space. Machine codes of all digital organisms, including the ancestor, are write-protected but are readable and executable by other organisms. In this sense, it can be said that organisms are single-cell creatures covered with a semi-permeable membrane. A mother cell, an ancestor program, makes its copy, a daughter cell. The daughter cell also makes its daughter, and then each individual cell repeats the replication independently in parallel until its death. An individual that fails to find space in memory and to replicate, or whose genome was rewritten so as not to self-replicate, dies out much faster. Such mortality is introduced by the reaper, which is a one-dimensional first-in first-out queue where all individuals are

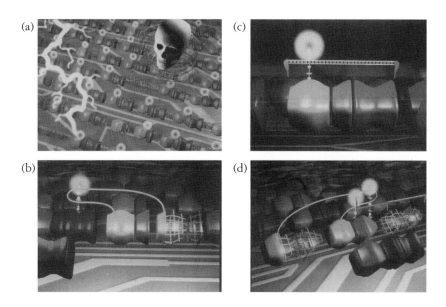

Figure 5.2 Visualisations of some of the main processes in the Tierra program.

registered and where their respective positions go up and down depending on the success or failure of executing machine instructions. The reaper is symbolically represented as the skull in the overall image of Tierra, Figure 5.2(a). Mutation is realised as random bit-flips with a low probability, which convert one to zero or zero to one in a machine code. There are two types of mutation: direct mutation by which a bit randomly chosen from the memory space is changed directly with the bit-flip, and copy errors, which occur with some probability in the reproduction process from a mother to a daughter.

In this virtual world that models mutation and natural selection, diverse and unexpected phenomena developed and were observed that surprised even Ray himself. First, ancestor programs filled the memory space as shown in Figure 5.3(a), and shortly afterwards a parasite appeared and increased as shown in Figure 5.3(b). Parasites do not have the copy procedure, the third gene of the ancestor, but can replicate more efficiently by borrowing the information of the copy procedure of an ancestor nearby, as shown in Figure 5.2(c). After a while, however, a program with immunity, called Hyper-parasite, appeared and became dominant, as shown in Figure 5.3(c) and (d). A Hyper-parasite, as shown in Figure 5.2(d), steals the CPU from a parasite and produces two daughters using the stolen CPU and its own CPU. Evolution continued beyond that stage, and thus ecological phenomena with much variation and diversity emerged that included, for example, host–parasite population cycles and sociality where several individuals form social cooperation to reproduce and so forth.

While such software evolution can be regarded as a process where new functionality and interactions are spontaneously and autonomously generated and diversified, another implication of Tierra is to prove the possibility of using such evolutionary mechanisms to optimise a specific function, i.e. a program or an algorithm. When focusing on a self-replication function given to the ancestor, the same function was eventually realised in one-fourth of the original code size and one-sixth of the original execution time. That is, the self-replicating program evolved itself to become as compact as possible in size. In other words, the digital organism could acquire the best survival strategy and could optimise itself in an environment where available energy and space are limited.

This shows us a new possibility of evolving a roughly designed program autonomously to an optimised one by using intentional or artificial selection to control the macroscopic conditions of the environment. This idea of macro control might also be extended to the possibility of indirectly controlling swarm or aggregation function by controlling environmental conditions. This is different from the conventional approach of directly controlling an individual.

These ideas of software evolution are expected to be a possible means of harnessing the evolutionary process for the production of complex computer

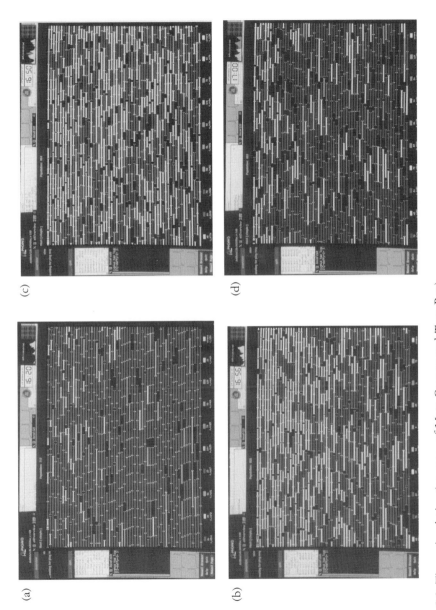

Figure 5.3 Tierra simulation (courtesy of Marc Cyguns and Tom Ray).

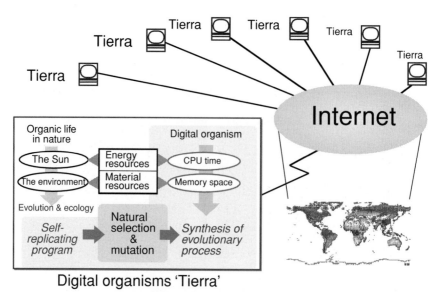

Digital organisms 'Tierra'

Figure 5.4 From 'Tierra' to 'Network Tierra'.

software. In order to investigate possibilities and limitations of software evolution, we have constructed a worldwide experimental system called 'Network Tierra' (Figure 5.4), where the evolutionary process is introduced into the context of massively parallel and networked computers to evolve complex Multiple Instruction Multiple Data (MIMD) parallel processes (Ray and Hart 1998).

Hardware evolution: Neural nets as hardware grows and evolves

Living organisms have adapted to their environments through a very long evolutionary process. Hardware evolution, or 'Evolvable Hardware', which enables the material structure of systems to evolve, is one of the most challenging and significant topics of research in the field of evolutionary computation (e.g. electronic circuits or molecular-based systems). Together with software evolution, it may enable the evolution of functional, controllable structures.

Figure 5.5 shows the framework of hardware evolution. Assuming sets of information as 'seeds' and spaces consisting of reconfigurable hardware devices as 'fields', what grows in the field from the seed is a digital circuit. Various hardware structures as digital circuits are generated on hardware spaces (fields) depending on sets of information (seeds). Every performance of generated circuits is evaluated based on the fitness function a human designer defines. Some good seeds that achieved much better performances than others

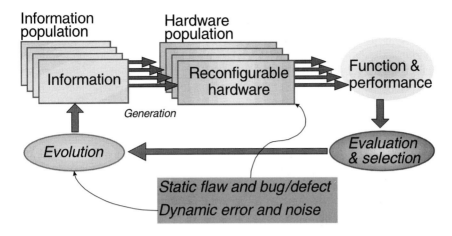

Figure 5.5 Framework of hardware evolution.

are selected, and the next generation of sets of information is reproduced by giving some genetic operations to the selected seeds. Repeating such generation and test, hardware structure is evolved generation-by-generation as the final product so that it meets the requirements for its function. Namely, hardware evolution is like a kind of breed improvement in hardware. FPGA (Field Programmable Gate Array) and Cellular Automata (CA) are typical reconfigurable hardware devices used as fields.

Hardware evolution in the form of electronic circuit evolution is very important and a practical research issue that we are now pioneering. As described above, human brains, especially infant brains, have very flexible hardware (i.e. wetware) structures. In an attempt towards building an artificial brain with comparable flexibility, we are conducting a project called 'CAM-Brain' (de Garis 1994) by using hardware spaces that three-dimensionally connect cellular automata as the fields.

Using special hardware called Cellular Automata Machines (CAMs) that provide three-dimensional Cellular Automata (CA) spaces in hardware, artificial neural networks are grown and evolved as models of nervous systems under the genetic control within the CAMs. Genetic information for growth is fed to the CAM sequentially, based on transition rules applied to all cellular automata cells. Cellular automata-based neurons extend trails of axons and dendrites, and synapses are formed when axons and dendrites collide, as shown in Figure 5.6.

Similar to biological brains, CAM-Brain is comprised of millions of interconnected neural modules. Each module has about a hundred neurons that are massively interconnected through their axonic and dendritic trees, with relatively few external connections to other modules. In order to evolve a module so that the module in a brain can perform a useful function, a

95

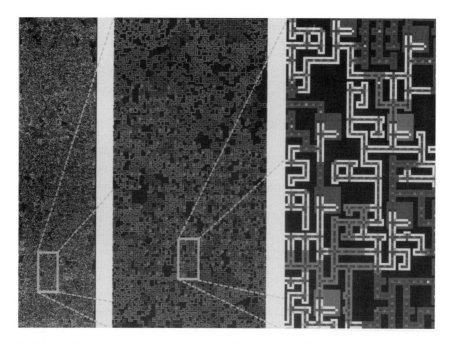

Figure 5.6 Neural network grown on two-dimensional cellular automata space.

genetic algorithm is applied; this algorithm is based on the Darwinian survival of the fittest through fitness-proportional reproduction. Typically, this involves a population of 100 modules evolving over 100 generations, which results in 10,000 module evaluations. Each module evaluation consists of growing a new set of axonic and dendritic connections in the three-dimensional CA space and then running it to evaluate the performance, which typically requires 100 update cycles for all of the CA cells in the module. Now, assuming that a module consists of 10,000 three-dimensional cells, as a result, a total of 10 billion cell updates is required to evolve a useful module.

In order to accelerate such brain building computations, we have developed a new brain evolution processor called CBM (CAM-Brain Machine) (de Garis *et al.* 2000). CBM can provide a 24 × 24 × 24 CA cell space in hardware by implementing CA with FPGAs, and can perform 100 billion cell updates a second.

A hardware evolution scheme has another significant impact on hardware manufacturing in the sense that hardware evolution is information-driven. Even if the same information as seed is given, if two different spaces are used as fields, the hardware structures grown would be different. This implies, in a sense, that each field does not have to be uniform, or that some manufacturing flaws in each field can be allowed, as is the case when a real field has some flaws like stones and/or holes. In other words, seeds that are simply appropriate for a given field would be selected and evolved through evolutionary processes.

Testability is said to be a serious problem in nano-electronics manufacturing, especially when enormous numbers of circuit elements can be placed on a single wafer. However, in the sense mentioned above, it will not be necessary to manufacture flaw-free devices. The use of evolutionary techniques to build flaw-tolerant electronic circuits would be a major breakthrough in the electronics industry.

CONCLUSIONS

Once a user goes into HyperReality, the user can interact with objects and communicate with people even if those objects and people are physically located in a different space anywhere on a space station or on Mars. That is, two users and their respective objects, belonging to two different physical spaces can virtually share a common space and world at the same time. In other words, the two spaces to which the two users and their objects belong are extended and merged into a common virtual space in HyperReality. In a sense, people and objects that physically exist can traverse real space to a virtual space.

On the other hand, another feature of HyperReality is that people and objects that do not exist physically but do exist virtually, without any binding to a real world, will also be able to show up in a physical world. Such entities are virtual characters that exist as robots in a physical world, which we discussed in the first part of this chapter. Such virtual characters/robots will be the artificial entities that benefit first from HyperReality.

In this chapter, virtual characters/robots in the near future have been envisaged from the viewpoint that such entities should play the important roles of attracting people into the HyperReality world and activating communication with people. The most significant characteristics required for such entities are autonomy and creativity.

In the remaining part of this chapter, we introduced the concept and research topics of evolutionary systems as the most effective means of creating an information processing system that is rich in autonomy and creativity.

A scheme of life-like and society-like information processing was proposed as a new engineering principle for evolutionary systems. Evolution and emergence and micro-macro dynamics are key concepts for evolutionary systems that will be able not only to develop new functionality in terms of software, but also to grow and evolve autonomously their own structure in terms of hardware.

The goals are to create evolutionary mechanisms to generate change, and to create emergent mechanisms to self-organise a system, based on the structure of micro-macro informational loops, as well as to seek and clarify various possibilities for artificial nervous systems using these mechanisms as evolutionary systems. Artificial brains that use these evolutionary technologies will be able to transcend the limits imposed on biological brains by their biological nature.

The ideas and approaches inspired by artificial life are increasing in importance and are expected to have a major impact on the areas of communication, robotics, and computer engineering. One of the most stimulating results in this respect has shown us the possibility of creating autonomy and creativity in computers. However, it seems that we do not yet have any methodologies that can be used to control such autonomy and creativity so as to meet engineering requirements.

How to achieve such controllability, which is contrary to autonomy and creativity, and how to adjust or merge controllability with autonomy and creativity, remain significant research issues, at least from an engineering viewpoint. It can be said, however, that in a sense we are now exploring a treasure trove of new research topics. It is likely that ALife-related research will be incorporated in such areas as sociology and economics as well as science and technology. The field of ALife itself will evolve and profoundly influence the development of HyperReality.

REFERENCES

de Garis, H. (1994) 'An artificial brain: ATR's CAM-brain project aims to build/evolve an artificial brain with a million neural net modules inside a trillion cell cellular automata machine', *New Generation Computing, OHMSHA LTD and Springer-Verlag*, 12: 215–221.

de Garis, H., Korkin, M., Gers, F., Nawa, N.E. and Hough, M. (2000) 'Building an artificial brain using an FPGA based CAM-Brain Machine', *Applied Mathematics and Computation*, 111: 163–192.

Langton, C.G., Tayler, C., Doyne Farmer, J. and Rasmussen, S. (eds) (1992) *Artificial Life II*, New York: Addison Wesley.

Ray, T. (1992) 'An approach to the synthesis of life', in C.G. Langton, C. Tayler, J. Doyne Farmer and S. Rasmussen (eds), *Artificial Life II*, New York: Addison Wesley.

Ray, T. and Hart, J. (1998) 'Evolution of differentiated multi-threaded digital organisms', *Artificial Life VI, Proceedings of 6th International Conference on Artificial Life*: 295–304.

Shimohara, K. (1994) 'Evolutionary systems for brain communications: towards an artificial brain', in R. Brooks and P. Maes (eds), *Artificial Life IV*, Cambridge, MA: MIT Press.

6

HYPERTRANSLATION

Minako O'Hagan

Editors' introduction
While surfing the Net we may be asked to select the language in which we want to read the pages of a web site. Or, we may be unaware that the site has recognised our country address and automatically selected an appropriate language. The Web is multilingual. Teletranslation is big business. The term 'teletranslation' was coined by Minako O'Hagan in her book on the subject (O'Hagan 1996). In this chapter, she moves on from teletranslation on the Internet to envisage what interactive language support will be like in HyperReality.

Terashima points out that synchronous communication in the coaction fields of HyperReality includes not only speech (the traditional concern of interpretation) but selected gestures and the deliberate use of body language. Minako O'Hagan uses the term HyperTranslation to suggest that translation and interpretation in HyperReality will take on new dimensions.

INTRODUCTION

As today's cyberspace develops into the infrastructure based on HyperReality (HR), in which we can be telepresent anywhere at any time and intermingle not only with real people but also with artificial creatures, our communication landscape will embrace a totally new dimension. William Gibson (1984), who coined the term 'cyberspace', creates a vision of a virtual person who can interact with a real person in physical environments:

> He looked into her eyes. What sort of computing power did it take to create something like this, something that looked back at you? He remembered phrases from Kuwayama's conversation with Rez: desiring machines, aggregates of subjective desire, an architecture of articulated longing. . . .
> (Gibson 1996: 237).

What Gibson depicts is the possibility of creating an artificial life form and its integration into the physical world. Sophisticated computer graphics combined with artificial intelligence make the creation of such a concept possible, and communications networks can bring them into contact with real people. Gibson's vision overlaps with the HyperWorld described by Terashima. If the current trends of progressive globalisation and the consequent need for localisation are any indication for the future, the Hyper-World can be expected to be a multicultural and a multilingual place in which people and artificial forms freely communicate and interact in a variety of modes irrespective of their language or cultural background. To this end, the HyperWorld with its new communications capabilities calls for a new generation of language support that functions effectively in the new environment.

This chapter considers the implications of HyperReality (HR) for the professional field of translation and interpretation (T&I). T&I refers to the facilitation of interlingual communication by converting one natural language into another, where translation deals with written texts in asynchronous mode and interpretation with spoken messages in an interactive mode. The author argues that new communications capabilities afforded by HR will drive the development of new modes of T&I.

During the 1990s the growth of the electronic global communications network has impacted on the translation industry, transforming it into a worldwide location-independent service. The concept of teletranslation (O'Hagan 1996) envisaged the wholesale shift of T&I to a service based on electronic networks, allowing translators and interpreters to be accessible in cyberspace. The World Wide Web has since provided a vehicle to deliver this concept, at least for translation. However, to date interpretation has largely remained location-dependent. With the advent of HR-based infrastructure, asynchronous and synchronous modes of language support for both text and speech are likely to be required in support of communication incorporating the virtual world. 'HyperTranslation' extends the concept of teletranslation to embrace the emerging duality of our communication space in three-dimensional, albeit virtual, presence-based communication. This therefore means that non-verbal communication elements will likely become a much more significant factor in language support in HR.

ERA OF DUALITY

The Internet has had a considerable impact on the translation industry, of which teletranslation now forms an integral part. The links between translators and their contracting companies or end-customers, as well as those between translators, are becoming more and more virtual in nature. This constitutes a shift from a transportation-based service to one based on electronic communications networks. Due to its predominant mode being asynchronous and text-based, translation has always tended to be a location-independent service, but the Internet environment has widened the scope of translation as a virtual service. Such shifts are taking place in many service industries. For example, while we can drive to a local music shop (bricks and mortar) to buy a CD, it is also possible to shop in a virtual store on the Web (clicks and delivery). Furthermore, the latter mode may have an additional feature based on an intelligent agent, which extrapolates your taste in music from your last purchase and makes recommendations from newly released titles. In this environment, the lack of a physical shopfront or a human shop assistant is irrelevant. The progressive migration of commerce into cyberspace suggests that T&I will have to be functional in virtual environments. Already translations can be purchased in cyberspace by a click of the mouse. In the near future, interpretation will also be required, not only for face-to-face gatherings, but also for virtual meetings. The new technological paradigm of HR will drive the need for interactive speech-based language support in addition to text translation functions. This will challenge interpretation practices that are grounded primarily in interactions in physical presence.

The Internet is progressively acquiring multimodal capabilities, such as voice communication enabled by Voice Over IP (VOIP) platforms, as well as image transmissions, in addition to its more established modes of e-mail and text chat. This makes it a supportive environment to host virtual meetings among people who are at separate locations. Over the last few decades meeting modes have been influenced by technology, beginning with the telephone. For example, today's face-to-face meetings may incorporate a telephone dial-in function enabling remotely located participants with a password to join in. Conference calls (audio conferencing) are routinely used by businesses as a means to link three or more parties on the phone and, while it has not become a mainstream mode of meeting, studio-based videoconferencing is also available. Internet-based computer conferencing is also used by some in the form of text chat. Rheingold (1995) has reported earlier examples of computer conferencing based on chat and mailing lists. In the context of language support applied to this environment, an early example can be seen with the Community Access '96 conference held in November 1996 in Nova Scotia, Canada (Ashworth 1996). In this conference, workshops and plenary sessions were broadcast online and the remote participants were able to interact with speakers and other conference attendees in real time via Internet Relay Chat (IRC). The conference had the special feature of online simultaneous interpretation of English and French, which was extended to off-site (virtual) participants who accessed the conference via computer. Conference papers and discussions were interpreted simultaneously for the live audience while, at the same time, typists keyed in the interpreters' words for the virtual participants who read them on a computer screen.

Although a number of computer-based translation systems are being developed for interlingual chat sessions for some language combinations, there are few reports of human-based counterparts. Aiming to explore how interlingual text chat sessions could be facilitated by a human interpreter, a pilot 'transterpreting' experiment was conducted by the Center for Interpretation and Translation Studies (CITS) at the University of Hawaii (Ashworth 1996). As the term transterpreting suggests, this is a hybrid activity that lies between today's text translation and interactive interpreting practice. The author was involved in a second similar experiment connecting four locations in Japan, the US and New Zealand with five participants and one 'transterpreter' who converted chat messages from Japanese into English in real time. The results indicated that it is possible for the human transterpreter to relay the message in real time by typing as she sight translates. Language support in this environment requires skills of sight translation of text as well as fast and accurate keyboarding. Furthermore, chat texts, which Herring (1996) calls 'interactive written discourse', are closer to spoken than written language, thus requiring transterpreters to be familiar with colloquialisms often typical of spoken communication. These characteristics of transterpreting are significant in relation to conventional T&I where the traditional demarcation of the task was based on whether the message was written or spoken, with transmission being asynchronous or synchronous respectively.

The duality of communication modes will mean a significant change in the working environments of translators and interpreters, as they will be required to cater to the needs arising from cyberspace. As a result, translators who have traditionally worked asynchronously on written text may find that they have to translate synchronously text that reads like a conversation, by typing in real time or in some cases using speech recognition. Similarly, the interpreter who traditionally worked in face-to-face situations may have to perform in distributed virtual meetings without actually seeing the participants who reside in different locations. The new environment not only affects the manner of delivery of the language support, but also the characteristics of the message with which T&I need to deal. The most prominent recent example is the translation (localisation) task of web sites. In this way, new language support will be shaped by the characteristics of the new communication media as is explored in the following section.

HYPERREALITY COMMUNICATION AND T&I

Any interaction in HR is mediated by computer and the coaction field can only be activated if a desired virtual setting exists in an HR database. Once the setting is chosen, all appropriate attributes to that particular setting are also provided, including objects and a knowledge database. Furthermore, HR interactions can only take place if participants have some common knowledge of a given topic (see Chapter 1, this volume). In this sense HR communications can be considered as highly contextualised and domain-specific (see Chapter 2, this volume). This characteristic has implications for T&I practice where domain-specific knowledge is a significant factor (Wilss 1996). For example, a translation of heart transplant procedures is better handled by specialised medical translators who are familiar with the subject field. Similarly, a genetic engineering conference will be more capably served if the interpreters have acquired a certain degree of understanding of the subject. In this sense, the knowledge database of a coaction field accessible to its participants will also be useful to translators and interpreters. Domain-specificity is even more significant with computerised translation systems. The main successes of machine translation (MT) systems for reasonably high-quality outputs have come from domain-specific applications, as in the case of the weather forecast translation system TAUM METEO (Nagao 1989). This is because, the narrower the domain of the communication, the easier it becomes to define the language use. This suggests that certain types of HR interlingual interactions within a strictly limited context may be served by some kind of MT.

HR has another feature that may influence the language support function. This is the capability of conveying non-verbal communication in an unprecedented manner. The emergence of computer-mediated communication (CMC) has highlighted the value of non-verbal communication because of the very lack of that element. The use of 'emoticons' as well as other images, icons and some innovative uses of language (such as capitalisation of letters to indicate shouting) was devised to compensate for this lack. With HR communication a new possibility emerges, not only to accommodate non-verbal communication, but also to be able to manipulate it. Biocca and Levy (1995: 148) suggest that 'the more sensory channels are supported by the more advanced virtual reality systems the more possibilities are open to both naturalism and augmentation'. This will be applicable to HR. For example, in HR environments the entire body of a person could be faithfully projected in telepresence in VR, but it could also be augmented in some extravagant ways such as displaying happiness by turning the whole body bright pink. However, in order to create a desired impact across different languages and cultural conditions, such manipulations will need to be part of recognised and established codes. The differences in various non-verbal communication cues from culture to culture, and the difficulty in establishing these, are well documented (e.g. Hall 1959; Argyle 1988; Dodd 1995). On the other hand, it is also possible that cyberspace may develop its own code of non-verbal communication that is universal to people who interact in the virtual world and beyond cultural differences.

Furthermore, non-verbal communication can be extended to a wider communication environment, given that every element in HR virtual environments needs to be built literally pixel by pixel. In this sense, they can be designed to accommodate more effective interlingual communication by providing opportunities for translators and interpreters to influence a wider context of communication. For example, an HR interlingual meeting place may be furnished with bicultural features reflecting the background of the participants or, if it is a business meeting, the seller may ask for the environment to be localised into the buyer's familiar cultural setting. It will also mean that the entire environment, other than immediate verbalised messages, will become part of the context of communication. Viaggio (1997) stresses the importance for interpreters of understanding the extralinguistic context via the non-verbal communication of both the speaker/listeners and the wider environment. In HR non-verbal communication can become much more explicit and therefore even more important than in the traditional modes of interpreting in primary reality.

A further characteristic of HR communication that is relevant to T&I is the artificial forms that can roam in the HyperWorld (see Chapter 1, this volume). They are computer programs today known as 'bots' or 'intelligent agents' that can learn the specific needs of the user and perform certain tasks by interacting with other such agents or information sources. Negroponte (1995: 102) maintains that 'what we today call "agent-based interfaces" will emerge as the dominant means by which computers and people talk with one another'. Although today's agents are mostly software invisible to the users, research is under way to incorporate avatar technologies with non-verbal communication cues in software agents (Lu *et al.* 1996; Noma and Badler 1997; Vilhjalmsson 1997). Such developments suggest that HR communication will take place, not only between humans, but also between humans and artificial forms that take a human appearance. As described by Turkle (1995), certain bots in some MUD (multi-user dungeon) and MOO (MUD object-oriented) environments are indistinguishable from humans in their interactions by way of written messages. A more recent example includes the world's first virtual newscaster Ananova (www.ananova.com) interviewing Kylie Minogue. Tiffin (Chapter 2, this volume) argues that one of the key factors of HR environments is that they provide a setting for human intelligence to interact with machine intelligence. In the context of language support function, machine intelligence may be used to translate texts written by humans and human interpreters may serve for artificial humans interacting with real humans.

TOWARDS HYPERTRANSLATION

Translation and interpretation are directly influenced by new developments of communications media as the language support needs to be functional in a given communications environment. For example, the emergence of teletranslation was instigated by the evolution of cyberspace, which enabled efficient electronic delivery of text. In this sense, the assumption that HyperReality will drive new modes of language support is logical.

According to Terashima, the motivation for creating the HyperWorld was to bring the current generation videoconferencing environment to something much closer to face-to-face interactions, particularly in terms of non-verbal communication such as eye contact. While this may reflect a particular emphasis on certain non-verbal cues in Japanese communication style, it is also significant in terms of language support function. In particular, interpreters who mostly work in face-to-face situations are sensitive to non-verbal cues of communicating parties. However, the processing of such information has been implicit rather than explicit in the conventional setting of interpretation. The HR environment, by comparison, allows the implicit nature of non-verbal information to become explicit. As discussed earlier, interpreters could design the meeting environment with consideration of the cultural implications for participants and manipulate non-verbal cues of the speakers to match the established codes within the target language culture. This, by far, is the most significant implication of the development of HR. Translation and interpretation have so far been dependent on verbal messages, but in HR translation and interpretation may become 'HyperTranslation', which makes flexible use of non-verbal elements in communication.

The following describes a future scenario for the language support function in an HR-based environment:

> Keiko is a Japanese/English cyber interpreter who specialises in environmental science and policy. She works freelance, based in Sydney, Australia and is under heavy demand, particularly because of her additional subject expertise. Most translation companies have highly integrated information management systems in place to handle a complex web of information and communication; every component of their work, which is normally carried out in separate worldwide locations, is controlled and managed by computers. They offer language management functions rather than isolated translations. Also, in order to respond to a wide range of demands, they have an extremely flexible company structure that allows them to form teams of experts for individual projects, and provides them with the necessary matrix of information according to job type. Freelance T&I specialists are furnished with an HR workstation with a datasuit. Keiko's workstation has a speech sensor and controls to adjust the speaker's voice to the optimum level (in case it is too faint, fast, high-pitched, etc.), a multimedia notepad for taking notes in a range of modalities, a stylus pen for writing on a distributed electronic virtual whiteboard, and a memory system that is tied to her own word list as well as recording her outputs. It also gives her access when she needs it to online help from her personal intelligent agent that she has trained, and to her human

interpreting partner. The datasuit enables tactile senses at a distance and also the control of the workstation functions. Furthermore the datasuit allows the interpreter's gestures to be superimposed on a specified image – normally that of the speaker whom the interpreter is serving.

Keiko's first appointment of the day is a hypermeeting of environmental scientists, each of whom is accessing the meeting from a separate location. In the hypermeeting the interpreters can share their clients' visual perspectives and tactile senses through their datasuits. For example, Keiko's perspective of the meeting can be switched to obtain her client's or the audience's visual perspective. She can also turn a limp handshake into a hard and crisp one as appropriate on behalf of her client. Preparations for interpreting assignments normally consist of checking background information supplied by clients and sometimes an advance meeting with them in HR. In this case, the intelligent agent of her client, Dr Sato, has sent Keiko's agent a password enabling access to the requested information. Through Dr Sato's hypermedia site Keiko was able to familiarise herself with his manner of speaking and key terminology by playing some of the memory system files. She was also able to refer to written materials, which were translated into a required set of languages for the meeting by a computer-based system. Most technical documentation destined for translation is now authored using controlled language, considerably improving the output translation quality.

The day arrives. As Dr Sato's presentation starts, Keiko's modulated male voice is heard speaking English. While presenting his paper, Dr Sato frequently pulls up a writing panel to illustrate a point, and Keiko uses her stylus pen to write any required translations of his text. Her English words appear beside Dr Sato's Japanese text on the panel in front of the delegates. Dr Sato's projected image on the screen is similar to his true self and yet his gestures are that of the interpreter's whose datasuit superimposes her non-verbal cues on to Dr Sato's image. This has resolved the problem of mismatched information experienced in earlier modes of conference interpreting many years ago where the audience watched the speaker's gestures but heard the interpreter's voice speaking their language. British audiences used to find it amusing to see highly animated Spanish speakers in front of them but to hear them speak in a way that was considerably toned down by the interpreter.

Keiko feels that the best part of hyperinterpreting is the increased control of the communication environment, including the client's non-verbal cues, as well as a realistic feel of 'being there'. Her cyberspace work mostly involves multimedia, making her task a hybrid of translating and interpreting. She particularly enjoys helping clients design their meeting environments. She feels that she is now able to make active use of her deeper cultural knowledge accumulated over years of living in both her native country and English-speaking countries. She certainly does not miss either the jet lag or the tense atmosphere of the interpreters' booth of traditional interpretation.

HyperTranslation as depicted above is a sophisticated form of distributed language support. Both the interpreter and each conference participant can access the meeting from different locations and carry on interactions as if they are gathered in one physical space. This environment has overcome the problems of the lack of visual cues with today's remote interpreting via telephone (Vidal 1998) or insufficient voice and image quality experienced in videoconference interpreting (Mouzourakis 1996). It has also overcome the mismatch of the listener's visual and audio information that occurs in today's face-to-face interpreting. It further provides new possibilities in terms of manipulating the entire communication environment. The HR environment also facilitates intelligent agent functionality, which could be combined with human-based language support systems. There will be a greater market for finely tuned services by human experts, while at the same time computers will provide effective aid to human experts as well as autonomous language support functions.

CONCLUSIONS AND FUTURE TASKS

Language needs are likely to become one of the essential requirements of mature information societies with global communications infrastructures. With HR technology a wide range of services will become available, allowing interactions with people or environments from different linguistic and cultural origins. Access to a service or information in the user's chosen language will become a prerequisite.

Consideration of the language issue has always tended to lag behind the creation and development of communication technologies. International telephone links were developed without any consideration of language barriers, and new developments in telecommunications technologies continue to reflect this attitude. Similarly, the focus of HR developments so far has been to overcome technical communications barriers with concerns such as making artificial sensory dimensions available for communication through the use of stereophonic sound, three-dimensional images, touch and smell. However, the implication of global communication networks has begun to dawn on communication service providers; telephone companies are providing telephone interpreting services while multilingual considerations are incorporated into today's popular browsers as well as search engines. As the global aspect of our communication becomes increasingly significant, the language issue will be further highlighted. What is more, computer-based translation systems are still not a satisfactory answer to this problem.

The initial failure of automatic translation systems stemmed from a belief that the translation task involved merely a quantitative shift of codes from one language to another. It took the MT technologists some time to discover that the multitude of languages used in human communication involved far more complexity than could be defined by mechanical dictionary look-ups. The limitation of computer application to translation is often described in terms of its lack of appreciation of context. MT systems typically analyse the message in terms of the 'objective' meaning stored in their dictionaries, whereas human translators and interpreters derive meaning dynamically according to the context. T&I require, in addition to linguistic knowledge, an understanding of extralinguistic meaning, which is supported by the context. It seems that machine intelligence, as we know it today, can only deal with disembodied meaning, whereas human intelligence is concerned with embodied meaning. Melby (1995) relates the latter to somatics – a perspective that accepts the involvement of the body in signification. Computer-based translation systems are unable to process information on the basis of multisensory functions as they are, in their current forms, blind, deaf and numb. Human translators and interpreters use all their senses in processing information. Such multisensory processing occurs even when dealing with written texts, which evoke images based on somatic experiences. While the gulf between machine and human currently seems irreconcilable in terms of achieving translation and interpretation tasks, the possibility of processing visual, tactile and olfactory information by computer may eventually furnish computers with multisensory functions. This may then give rise to the possibility of apparent understanding of a given communicative situation by artificial intelligence, which effectively functions in multimodal environments such as in HR. However, given the history of MT research of the last half century and that of AI, the likelihood of such a system within the foreseeable future is at best uncertain.

A sophisticated communications environment such as HR, where we can be represented in any forms we wish in telepresence, will render even more acute communication barriers due to the differences in language and culture. The HyperWorld could be an exciting place for translators and interpreters while at the same time the new communications environment will undoubtedly pose significant new challenges. With its promise to enable the seamless union of virtual and physical space, HyperReality defies limits to our communication over distance that were once considered impossible to overcome. Flexible language support that can extend beyond the verbal message to intricate non-verbal communication will complete the picture of a truly global communications infrastructure.

REFERENCES

Argyle, M. (1988) *Bodily Communication*, second edn, London: Methuen.

Ashworth, D. (1996) 'Transterpreting: a new modality for interpreting on the Internet', Pan-Pacific Distance Learning Association Conference, Hawaii, PPDLA.

Biocca, F. and Levy, M. (eds) (1995) *Communication in the Age of Virtual Reality*, Hillsdale, NJ: Laurence Erlbaum Associates.

Dodd, C.H. (1995) *Dynamics of Intercultural Communication*, fourth edn, Madison, WI: WCB Brown & Benchmark Publishers.

Gibson, W. (1984) *Neuromancer*, London: Victor Gollancz.

Gibson, W. (1996) *Idoru*, New York: G.P. Putnam.

Hall, E.T. (1959) *The Silent Language*, Greenwich, CT: Fawcett Books.

Herring, S. (ed.) (1996) *Computer-Mediated Communication: Linguistics, Social and Cross-cultural Perspectives*, Amsterdam: John Benjamins.

Lu, S., Yoshizaka, S. and Miyai, H. (1996) 'A human-like computer character user interface by creating communicational gestures', *NEC Research and Development Journal*, 37, 2: 275–290.

Melby, A. (1995) *The Possibility of Language*, Amsterdam: John Benjamins.

Mouzourakis, P. (1996) 'Videoconferencing: techniques and challenges', *Interpreting*, 1, 1: 21–38.

Nagao, M. (1989) *Machine Translation: How Far Can It Go?*, trans. N. Cook, Oxford: Oxford University Press. (Original work published 1986.)

Negroponte, N. (1995) *Being Digital*, New York: Knopf.

Noma, T. and Badler, N.I. (1997) 'A virtual human presenter', IJCAI-97 Workshop on Animated Interface Agents, Nagoya: IJCAI.

O'Hagan, M. (1996) *The Coming Industry of Teletranslation*, Clevedon: Multilingual Matters.

Rheingold, H. (1995) *The Virtual Community*, London: Minerva.

Turkle, S. (1995) *Life on the Screen: Identity in the Age of the Internet*, New York: Simon & Schuster.

Viaggio, S. (1997) 'Kinesics and the simultaneous interpreter: the advantages of listening with one's eyes and speaking with one's body', in F. Poyatos (ed.), *Nonverbal Communication and Translation*, Amsterdam: John Benjamins.

Vidal, M. (1998) 'Telephone interpreting: technological advance or due process impediment?', *Proteus*, 7. [This is the digital edition of the newsletter of the National Association of Judiciary Interpreters and Translators.] Available at http://www.najit.org/proteus/vidal3.html

Vilhjalmsson, H.H. (1997) '*Autonomous Communicative Behaviours in Avatars*', MS thesis, Massachusetts Institute of Technology.

Wilss, W. (1996) *Knowledge and Skills in Translator Behaviour*, Amsterdam: John Benjamins.

7

THE HYPERCLASS

John Tiffin and Lalita Rajasingham

Editors' introduction

In December 1997 John Tiffin, Lalita Rajasingham and Nobuyoshi Terashima began an action research project to implement a HyperClass on the Internet. Essentially this has meant a series of test transmissions with consequent development of the technology that has established the technical viability of a HyperClass on the Internet. In parallel to this they are developing a pedagogical design process for the HC based on a neo-Vygotskian approach to education as a communication process. They anticipate that education will be the first major application of HR and aim to have the first HyperCourses in a HyperClass setting offered as part of a Master's course in a HyperUniversity in the early years of the new millennium.

HYPERCLASS – THE NEED

Domingo Sarmiento said 'There is only one problem and that is education, all other problems are dependent on this one.' Sarmiento was the President who founded the educational system of Argentina. He wanted to make Argentina a modern industrial nation. National education systems have come into existence over the last hundred years in response to the needs of commerce and industry in industrial societies. Today, on the Internet, commerce follows education. It is schools and universities that have pioneered the use of the Internet. It is the young that lead the way. Students surf the Net before their teachers and it is often the school principal who has least understanding of cyberspace. To know where the future is going, it may be that we should be looking over the shoulders of our children as they seek it out.

We watched a group of New Zealand students join a group of Japanese students for a seminar. The medium was videoconferencing. Both students and academics saw their counterparts in the other country as two-dimensional images on a video monitor. The seminar proceeded in a conventional manner. Papers were presented by academics from both countries and there were questions from students. Pictures were carefully composed by video cameras and framed on the monitors in the manner of a television programme.

110

At one point there was a break of 15 minutes and the videoconferencing link was left open so that the students in the two countries could communicate socially. Suddenly the communication changed. It seemed as though the video monitor had become a dormitory window through which students were leaning and chatting with frank curiosity about each other: 'Hi, what's your name? What subjects are you taking? How about a date?' One felt the raw desire for a technology that would allow students to climb through that window and freely intermingle as full-bodied three-dimensional beings. This is the basic goal of a HyperClass, which we are currently pursuing with Dr Terashima in a three-year experimental project (Terashima, Tiffin and Rajasingham 1999).

In 1994, when we were writing *In Search of the Virtual Class: Education in an Information Society*, we made the basic case that a new generation of telecommunications is coming into place. It will provide the infrastructure of an information society, where people depend more on telecommunications and less on transport than is the case in an industrial society. So an educational system is needed that will prepare people for life in such a society. We saw the current use of telecommunications in education – for e-mailing, bulletin boards, conferencing, generating web pages, surfing and videoconferencing – as steps along the road towards a virtual class which would ultimately be in distributed virtual reality on the kind of broadband networks that fibre optics promised. When we saw the work being done by Dr Terashima at ATR we realised that the virtual class will become a HyperClass.

Education as we know it is a preparation for life in a nation state. But we are in the process of becoming a global society and we need an education that prepares us for life in a global society. We need global education, not instead of, but as well as, national education.

Education today takes place in classrooms and schools as a preparation for life in the rooms and buildings of industrial societies. But we are becoming an information society, an anywhere anytime society that, since the coming of the Internet, lives on the cusp between global cyberspace and local realities. If HyperReality is to be an infrastructure of the information society that makes this possible, then we need education in HyperReality.

HYPERCLASS – THE CONCEPT

An educational system is a communication system that brings together the four critical components that are needed for learning to take place. These are teachers and learners, knowledge and problems (Tiffin and Rajasingham 1995). The core communication process permits teachers to help learners to apply knowledge to problems and is called a class (Figure 7.1).

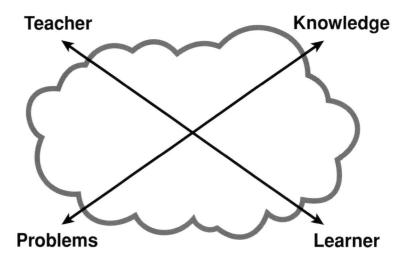

Figure 7.1 A class is a communication system that allows teachers to help learners to apply knowledge to problems.

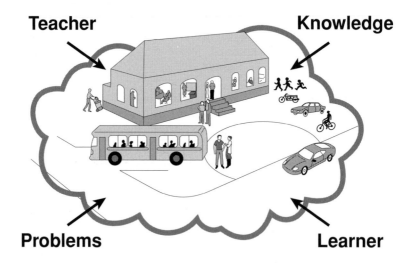

Figure 7.2 A conventional class is a communication system that uses transport systems to bring teachers and students together in a classroom where teachers can help learners to apply knowledge to problems.

It is transport systems that bring these four factors together in conventional educational systems. And it is buildings and, in particular, the classrooms within them, that provide secure space where interaction can take place between teachers and students about the application of knowledge to problems. Today's schools, colleges and universities could not exist

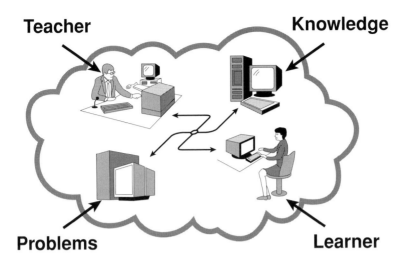

Teacher

Knowledge

Problems

Learner

Figure 7.3 A virtual class is a communications system that uses telecommunications and computers (the Internet) to allow teachers to help learners to apply knowledge to problems.

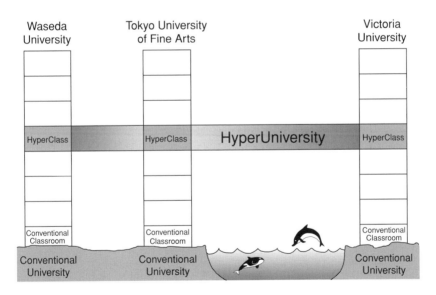

Figure 7.4 HyperClasses in HyperUniversities.

without the roads that make it possible for students and teachers to come together to make use of the buildings that house libraries and classrooms and support systems (Figure 7.2).

Virtual schools, colleges and universities are appearing on the Internet. What distinguishes them is that they use computers and telecommunications instead of buildings and transport to bring teachers, students, knowledge and problems together in a virtual class as bits of information rather than as atomic substance (Figure 7.3).

A HyperClass is a class that is conducted in an environment where physical reality and virtual reality commingle in a way that becomes increasingly seamless. It is a combination of a conventional class and a virtual class. Teachers, students, knowledge and problems can come together in a class via local transport and they can also come together via the Internet.

HyperSchools, HyperColleges and HyperUniversities can exist in real and virtual dimensions at the same time. In so doing they will provide an intersection between the local and global dimension in education. This means that a HyperClass is necessarily a synchronous activity, whereas a virtual class can be asynchronous.

In Figure 7.4 the vertical blocks represent the buildings of existing universities. The sections in them represent real classrooms/lecture theatres/seminar rooms. A stretch of ocean indicates that the universities can be in different countries. The horizontal shaded blocks represent a virtual dimension and the possibility of linking classes in different universities in different countries in a HyperUniversity. A student could go to a conventional class in a conventional university or stay at home and use a PC and the Internet to link to a virtual class in a virtual university. A HyperClass allows a student to do both. A HyperClass exists where the virtual and real dimensions intersect. This is a coaction field where students and teachers in a conventional classroom can synchronously interact with students and teachers in other universities that may be in other countries. The current experimentation seeks to make this link between a class in Waseda University, Japan and a class in Victoria University of Wellington (Terashima, Tiffin and Rajasingham 1999). The same diagrammatic explanation could be applied to HyperSchools and HyperColleges.

THE TEACHER–STUDENT AXIS IN THE HYPERCLASS

The relationship of the four critical factors of education can be thought of as the intersection of two axes: the teacher–student axis and the knowledge–problem axis. The fundamental concept of teachers helping students to learn how to apply knowledge to problems is derived from Lev Vygotsky's (1978) concept of a zone of proximal development (ZPD).

Virtual educational institutions that are developing on the Internet essentially address self-sufficient learners who can study by themselves and are satisfied with a course of study that is stripped down to the cognitive essentials. The conventional classroom by contrast is a social setting for learning where, besides the institutionally recognised learning agenda, there is unofficial social interaction. We have been conducting action research on the virtual class since 1987 (Rajasingham 1999; Tiffin 1990a; 1994; 1998) and we consistently note that, while people can study at home on their own, many want to learn as a group. While it seems to make little difference, as far as purely cognitive skills are concerned, whether students are in a conventional classroom or using a PC and telecommunications from their home or office, our telestudents go to great lengths to meet with each other for the social interaction that goes with being part of a class. They want to gossip and find out how fellow students feel towards what they are learning and how the course of study is affecting their lives, and they want to do this in a face-to-face situation where they can note what other people look like as they speak, how they move, and what they wear. They want to eyeball each other and their teachers in the round. A HyperClass is a coaction field that allows real and virtual teachers and students to interact together for learning. In doing this it would seem to have more potential, through the use of avatars, to create a sense of group membership between real and virtual people than is possible with conventional audio- or videoconferencing. However, it is unlikely in the foreseeable future that it will fully satisfy the human need for real human interaction that seems to be a concomitant to learning for many people.

Participation in a real school, college or university can be thought of as a vertical integration of people from the local hinterland that the institution has traditionally served. People who live in the same locality join together for the social and sporting activities associated with education and for subjects that are local in nature or have social, physical or practical components, for example, where people study their native language, learn to sing and dance, play games, do sports or study local history and geography. By contrast, participation in a virtual college can be seen as a lateral linking of people who share the same interest in a subject, or where the subject is of a global nature or where people seek to study outside the cultural confines of where they live. People can choose to be of any race, religion, gender or age when they come together in a virtual class. It is a good place for learning foreign languages, science, economics, global geography and global communications.

In one dimension, an arts student can date a science student and a computer scientist play tennis with someone studying marine biology. In the other dimension, an elderly woman in Budapest can share her passion for Gothic architecture with a young boy in Rio de Janeiro and an accountant in Tokyo.

The HyperClass is where the virtual and real dimensions of students and teachers intersect. It provides a field for reconciling local learning with learning that is global and for understanding a subject from the multiple perspectives of cultures other than one's own.

The association in a HyperClass of people who have come together because of their interests in a specific subject suggests the development of

HyperColleges dedicated to specific disciplines: in contrast to the Universities of Madrid, Tokyo and Berlin, whose very names emphasise location. We would see the emergence of HyperUniversities of Communications, Nanotechnology and Hindu Literature. This fits with Terashima's definition of HyperWorlds as technical environments where coaction between reality and virtual reality is based on a shared domain of knowledge.

At its most basic, the HyperClass is the kind of situation envisaged at the beginning of the chapter, where a class of students in one country teleconference with a class in another country or indeed in several other countries. The difference is that, instead of being framed in a two-dimensional video display unit, the people in the class appear to each other as three-dimensional, life-sized human beings apparently occupying the same space. The walls of a classroom in one place disappear to reveal classes in other places in other countries. One of the defining attributes of the modern nation state is a national educational system and a national library. The national library houses a body of knowledge about the country, its geography, its history, its literature, music and arts, its culture and its language. The national education system will, at a primary, secondary, and tertiary level, seek to teach this body of knowledge in the language of the nation to those who will become its citizens. The national education system will also teach numeracy and literacy and bodies of knowledge such as science and mathematics that are global in nature.

Go into any classroom anywhere in the world and there will be a single teacher in front of a body of students using blackboards/whiteboards, textbooks and exercise books to teach people how to apply knowledge to the problems they and their country face. Much of what is taught is the same in any country, but the application of the knowledge is seen in terms of the needs, the nature and the mythology of the nation that is paying for the education system through its taxation system. Students grow up with a sense of nationhood and of social solidarity with their fellow students. However, the growth of a global economy and a global workforce sees the need for a global education system that prepares people to be part of that global workforce, able to act as global citizens and with the knowledge to cope with global problems. Systems based on HyperReality, unlike systems based on transport, do not recognise national boundaries. It becomes possible to hold classes and study the application of knowledge without regard for national borders. A class of Inuit in Alaska can study the Inuit language, culture and traditions with a class of Inuit in Russia. Peoples who have had diasporas such as the Jews, the Tamils and the Celts can maintain their cultural networks. Managers in multinationals can do an MBA that looks at legal, economic and managerial conditions across countries and, by learning with people in different countries, they can create the global networks they will work with in the future.

It is not just students from different countries who can meet in HyperReality. Teachers from different cultures can meet in a HyperClass as well. In consequence, the HyperClass need not be dominated by the perspective of a single teacher. Knowledge and its applications can be understood from a multiplicity of perspectives. One of the serendipitous outcomes of our research over the period since 1994, when we became Internet-based, was the ability the Net provided to almost casually bring into a class experts from many countries and fields. Students even began using the e-mail to ask questions directly of the authors of their textbooks and to bring outside authorities into a seminar they were presenting.

Unfortunately human communication problems do not disappear on the Net. Having a global dimension to one's classes exposes one to the dangers as well as the joys of intercultural communications. We have had flame wars between students of different cultures as well as acts of diplomatic sensibility from students that would provide a model for the United Nations. Perhaps because of the strong traditions that surround the idea of a teacher being responsible for a class, problems are not widely reported. When we mention at conferences that we have had flame wars in using the Internet we invariably have people coming up afterwards to share similar problems. Not surprisingly, some teachers in a virtual class situation find the unpredictability of the classroom dynamics using the Internet daunting. We can expect that the HyperClass will generate its own communication problems.

So far we have looked at the teacher–learner axis in terms of the real and virtual dimensions of a HyperClass, noting that they tend to serve different aspects of education, but that these can be seen as complementing each other to provide a whole education for a person who wants to be a citizen at the global and the local level. We need now to turn to the fact that the axis also functions synchronously and asynchronously and at different fractal levels.

A conventional class is a synchronous communication process that shifts between different levels of communication. As teachers seek to help learners to apply knowledge to problems, they may interact with the class as a whole, with a group of students or with an individual learner. Explaining to a class as a whole the teacher may ask, 'How many of you think that is the right answer?' and then address the group who responds. A hand goes up and the teacher now addresses an individual, then turns the question from the individual back to the class as a whole. The interplay between different levels can shift rapidly in complex patterns including, on occasion, students taking on the role of teacher.

Such shifting of communication levels is not easy to obtain synchronously in the kind of virtual class that obtains today with videoconferencing. It has to be organised. Cameras and microphones have to be oriented to individuals and it is difficult to do this effectively with large groups. Advance planning is needed for points at which a class will work together as a whole and points at which people will work in groups or pairs. Trying to create a virtual class through videoconferencing on the Internet has proved difficult.

Until it is fully linked in broadband networks, the Internet is not a truly synchronous medium and there are variable time lags between responses that make it difficult to interact naturally. Mixing real and virtual classes in a HyperClass at different levels has a long way to go before the process becomes as spontaneous as it is in a conventional class.

The upper level of a conventional class is restricted by the size of class-room in which it operates. The virtual class and hence the HyperClass do not have these restrictions. Some schools do have systems where classroom walls can be pushed back to allow several classes to be joined together for large group activities such as a school assembly. HyperClass experiments so far have linked two classes – one at Waseda University in Japan and the other at Victoria University of Wellington in New Zealand. Soon it will be possible to link six classes at six different universities. Ultimately there are no limits to the number of students and teachers who can be linked. It makes possible a new fractal level for the class – that of the superclass.

In theory a superclass could number millions, resembling something like the masterclass lectures we see on television where a gifted teacher broad-casts to a much larger 'class' than the one that is assembled in front of them as a form of studio audience. The difference, however, in an HC super-class is that it would be part of an integrated learning system that included large numbers of classes, groups and individuals, so that it would be possible to shift fractal levels before, after, or even during a class. Imagine the teacher asking, at the superclass level, 'Can any class tell me . . .?' and, on getting a response from a class in Madras, having that class join him at that moment so that he could interact directly with them. This could then lead to an interchange with an individual student in that class. We have to recognise that in such superclasses the role of teacher would operate at many different levels and would involve many people. The teacher–learner axis would no longer involve one teacher with 30 or 40 students but could well require teachers at tutorial level, group level and class level as well as different teachers for different parts of a superclass. We have for some time been moving towards a dream-team approach to lecturing where we include different lecturers, each with a speciality for different parts of a lecture, with the course lecturer acting as a coordinator, linking the parts to the programme as a whole. Conventional classes not only switch between different levels of communication, they can also switch from synchronous to asynchronous mode. Few of us will forget the exhortation to take out our textbook and turn to a particular page. It switched a class from synchro-nous to asynchronous mode. In many education systems the asynchronous mode is closely associated with some form of exercise to see if the learner can apply a certain field of knowledge to a certain domain of problems. This is often associated with a norm-referenced marking system that promotes competition. As a result, peer or group communication may be discouraged and regarded as cheating. By contrast, the virtual courses that have developed on the Internet in the late 1990s seem to be essentially asynchronous in nature and to promote peer and group work on assignments.

The HC is, like a conventional class, essentially synchronous in nature and, like a conventional class, can readily embed asynchronous episodes during the class as well as before and after it. However, to prepare people for life in an information society where networking is a valuable skill, the HC will need to facilitate communication at all levels, in synchronous or asynchronous mode and between virtual and real components.

To participate in a HyperClass students and teachers will need avatars. One can see some interesting arguments along the lines of whether or not to wear school uniforms in HC. Should students use official school avatars for first- to fourth-formers? Can prefects wear avatars of their own choice? Do they have to resemble the actual person and, if so, how much? Can students cross-dress? Moving through such Internet VRML sites as WorldChat or Black Sun suggests that we have a lot to learn about communication protocols between virtual people if we are to move this kind of communication beyond the level of street graffiti.

Perhaps the most profound aspect of the teacher–learner axis in a HyperClass is that the avatars of teachers and students may not necessarily represent human intelligence. HR is a platform for human intelligence to interact with artificial intelligence. As this is being word-processed, a little cartoon character looking like a paper clip keeps popping up on the screen to say it sees someone is trying to write a letter and could it help. Someone has tried to program a just-in-time (JIT) artificially intelligent teacher that follows Vygotsky's ZPD model of learning. It detects that someone has a problem and comes along to try to help them apply the knowledge they need to deal with it. The fact that the device pops up when there is no problem and offers the wrong help and never turns up with useful knowledge when there is a problem does not mean that one day such wizards won't work. The idea is already there for JIT teaching agents, especially when the instructional tasks are clear and there are strong patterns of student needs and frequently asked questions (FAQs).

Marvin Minsky (1986) first proposed the idea of small programs called agents with a degree of autonomy, an appearance of intelligence and the ability to work collaboratively. Interlinking agents are now being devised for office systems to look after such tasks as schedule management, e-mail management, meeting arrangement and workflow management (Asakura *et al*. 1999). A HyperClass needs to reflect the modus operandi of the society it prepares people for. It makes sense to have agents who will set up a HyperClass and ensure that everyone and everything is present. Already the management of e-mail and searching for information on the Web have become major problems for teachers using the Internet. Some of the problems noted above for shifting communication levels in a HyperClass could be alleviated by a communication flow management agent.

THE KNOWLEDGE–PROBLEM AXIS IN THE HYPERCLASS

Every society develops bodies of knowledge that explain and deal with the problems its people face. Whether it is in temple carvings, oral traditions or libraries, these bodies of knowledge exist independently of the human beings who helped create them. The second axis of learning involves the relationship between such bodies of established knowledge and the problem domains they address. The way this relationship could be addressed in a HyperClass suggests yet another profound contrast to conventional class-room processes.

A conventional class normally studies the application of knowledge in symbolic form to problems that can be expressed symbolically. For example, knowledge of how to write can be applied to the problem of writing a letter or an essay, knowledge of mathematics to problems expressed in mathematical notation. To the extent that the subject matter of a HyperClass is essentially alphanumeric and requires no more visualisation than can be provided by diagrams and two-dimensional still pictures, all that is required of a HyperClass, besides the means for real and virtual students and teachers to interact, is some kind of virtual whiteboard/blackboard. This needs to be readily accessible to everyone in a tutorial or a seminar and to the teacher in a lecture situation. The whiteboard acts as the short-term memory of an instructional event. Regardless of whether it is in a real classroom or a virtual class, a whiteboard is one of the most basic and powerful instructional devices at a teacher's disposal and needs to be available in the HyperClass.

Books are the long-term memories of conventional classes and students usually have a number of texts to hand that they can quickly refer to. School libraries serve as anchors to classroom activities. The HC will need to have the equivalent of books and libraries and some way for students and teachers to call up a page in a form that is readily readable. Since its invention in 1990 the Web has become the biggest library the world has ever known. However, with over 40 billion pages to search, finding things is becoming a serious problem (Takano and Kubo 1999) and advanced forms of Web crawler for HyperClass-based education will be critical.

When problems have a real-life referent, classrooms with whiteboards may not be the best place for learning. For example, learning to apply geographical knowledge, medical knowledge or knowledge of how to drive a car to problems presented in alphanumeric or diagrammatic form on a classroom whiteboard in a classroom is very limiting. Stories of the consequences are legion: students who pass a multiple-choice question on what a volcano is, but cannot recognise that their school stands on the slopes of one; medical students who can write an essay on a disease, but cannot recognise the symptoms in a patient; people driving away after passing the written driving test and failing to observe a rule of the road that they had just correctly given on paper. This is the problem of the transfer of learning from classrooms to real-life situations. It is a problem that is seldom addressed because of the way knowledge learned in classrooms is tested in classrooms. Solutions to problems are examined in the way they were learned, alphanumerically. This is quite different from testing the application of knowledge to real-life situations in those real-life situations.

The HyperClass introduces a new dimension to education by directly juxtaposing knowledge with the kind of problems that have a referent in physical reality. It is not easy to take someone living in Britain on a field trip to a volcano, and to ensure that they can, in safety, see the volcano erupt. It is difficult to present malarial students in delirium to students at medical schools in Japan. Most driving schools teach within the reality of the roads on which the student is likely to be tested, but they do so in safe conditions and try to avoid the nightmare freak conditions that one day the learner driver is going to have to face. HyperReality can take a class to an active volcano, to the bedside of a patient in crisis or to a critical traffic situation. At the same time it can window key knowledge that students should have in mind while they study the problem and try to solve it.

The means to do this lie in the application of the capability described by Ahuja and Sull (Chapter 3 this volume) to develop a three-dimensional virtual reality in one place from images garnered by cameras in another place. It allows the development of virtual reality simulacra of case studies from a problem domain. The eruption at Mount Ruapehu mentioned in Chapter 8, this volume, was recorded by an array of video cameras at different sites and these could be combined and converted into a virtual reality simulacrum of a volcano erupting that could be a case study for future HyperClasses. Medical students in North America could accompany a doctor in Africa on her rounds through a malaria ward. Observations of malaria patients could be related to the kind of four-dimensional human atlas we already find on the Internet, that allows us to view the functioning of organs in 3D. It is common practice to have video cameras permanently stationed at dangerous traffic intersections. These could be used to create VR libraries of problem case studies that include dangerous conditions. A learner driver could then drive a virtual car through a virtual reality of a specific intersection in which they could be faced with a vast array of different situations.

Where students are faced with practical problems in today's education and training systems, their comprehension of the problem domain is limited by their perspective of time and space. To a human, a volcano is something very big whose formation and existence is a matter of millions of years. By contrast, a malarial parasite is too small for a human to see and its life cycle is brief. Motorists may sense the size and speed of their vehicle in relation to that of other road users, but they cannot see the way they use the road from the perspective of other drivers and pedestrians. In a

HyperClass it would be possible to augment reality by switching view-points with another driver or by taking a helicopter view of the section of road that presents the particular problem under study. Similarly it would be possible to have a fly-over virtual reality of a volcano, or to window a satellite view of the extent of an eruption, or compress the timescale of the volcano's history so that it can be viewed as it changes shape with each eruption. It would be possible to compress the history of a patient with malaria and compare appearances at different times, or to take a microper-spective of what was happening in their blood during a delirium.

This ability to collect case studies as virtual realities that can be studied from multiple perspectives suggests a new methodology for research. Typically today's researcher conducts an experiment to test the relationship between a theory and the problem domain it addresses. This is normally done by sampling the problem domain, applying a treatment to a problem on the basis of the theory and reporting in words and numbers. Those aspects of a problem that cannot be measured and described alphanumerically tend to be ignored. The view of the physical reality of problems that trickles down from researchers into the bodies of knowledge that are taught in education and training systems is primarily embedded in words and numbers. While that part which is in numbers may be considered objective in what it measures, it is also selective in the aspects of the problem considered.

Where research is reported in words every reader of the research creates a different mental image of the physical referent. This is compounded when research is based on the publication of previous research. The extensive bibliographies so highly esteemed in published research are there because the reading of research is based on interpretations of interpretations. VR simulacra of problems model the original referent of the problem and are tied directly to it. In its educational use everyone is observing the same thing, and it is readily identifiable with the reality it is based on.

Simulacra developed in research can be utilised directly in HyperClasses in the manner in which a video might be used in a conventional classroom to illustrate an issue. However, where the simulacra or the virtual repre-sentation of some phenomenon forms the basic subject of a course, then it may impose a structure on the course. It is possible to think of a professor of anatomy holding her classes in her course on cardiology in a human heart. It suggests the development of HyperCourses for HyperClasses.

THE ESSENTIALS OF COMMUNICATION IN A HYPERCLASS

Figure 7.5 is an attempt to depict the different aspects and elements of a HyperClass as a pedagogical communication system. A teacher and a learner can be virtual or real. A virtual teacher can have human intelligence (HI)

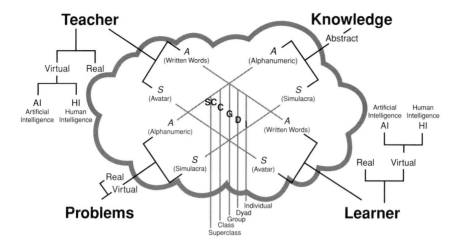

Figure 7.5 Communications in a HyperClass.

or artificial intelligence (AI). Teachers and learners can communicate synchronously *(S)*, using speaking avatars, or asynchronously *(A)*, using written words. Knowledge and problems can be embodied in the teacher and the learner or they can be represented alphanumerically or in simulacra. Problems can be real, however knowledge is always abstract. The interaction of teachers and students over the application of knowledge to problems takes place at the level of the individual (I), the dyad (D) the group (G) the class (C) and the superclass (SC).

THE COMING OF THE HYPERCLASS

National educational systems are paid for with taxes. The cost of building schools and colleges and the roads that lead to them increases, as does the cost of paper for textbooks and exercise books and the training of teachers. Costs of conventional education are incremental. Every additional 30 or 40 students means an additional room, teacher and infrastructure. By contrast, the cost of using information technology decreases, while computer bandwidth and capability increase. Information technology-based education depends on market size. The more people use the technology, the less it costs per head. A lecture can as readily be given to a billion students as to a single person. Compare the cost of a computer to the cost of a car. Compare Internet transmission costs to the costs of petrol and parking. Consider the costs of buildings and the grounds they occupy. Remember that transportation systems, unlike telecommunications systems, are subsidised by governments through taxation, and that it is the trend today for countries

123

to reduce taxation by seeking to reduce their contribution to education. In an age where market forces dominate, it becomes increasingly apparent that it costs less to bring the critical components of education together by using telecommunications instead of public transport and computers instead of buildings.

This could mean the commercialisation of education and the development of global educational corporations that trade in teaching (Tiffin 1990b). Does this mean the end of the conventional bricks and mortar classroom? Although teleworking, teleshopping and other teleservices could reduce our dependency on offices and shops and workrooms, there is nothing to suggest that they will make the use of rooms obsolete. Rooms are an ancient technology basic to all known civilisations. It is unlikely that we will stop using rooms in an information society and, if we are to continue to prepare people for life in rooms, then we need to teach them in rooms. Schools, moreover, have a custodial function. Learning at home and alone is not for the very young.

One view of the information society suggests the growth of telecottages or smart houses where rooms can be used for teleworking, telelearning and other tele-activities. In industrial societies the walls of rooms have an electrical infrastructure that provides power for lighting, heating and appliances. The walls of rooms in an information society will be networked for information services that could include a whole range of instructional services.

The conventional classroom will still be needed but education has to adapt to the virtual dimension of the information society. The HyperClass interrelates the two modes. Sometime in the early years of the third millennium the technology for HyperClasses or something like them will become widely available. At first it will be circumscribed by current PC technology and as primitive as the first attempts to fly or to use radio for communications, but it will improve. Restrictions on its development will lie in the mindset of traditional educationalists and perhaps even more in the attitude of educational administrators who are more truly threatened by these changes. Perhaps the answer lies in the way students seem to be the adopters and teachers the followers with the technology of the information society.

REFERENCES

Asakura, T., Ishiguru, Y., Kida, K., Yoshifu, K. and Shiroshima, T. (1999) 'Office work support with multiple agents', *NEC Research and Development*, 40, 3: 308–313.

Minsky, M. (1986) *The Society of Mind*, New York: Simon & Schuster.

Rajasingham, L. (1999) 'Trends in the virtual delivery of education: New Zealand and Pacific Islands', in G. Farrell (ed.), *The Development of Virtual Education: A Global Perspective*, Vancouver: Commonwealth of Learning Press.

Takano, H. and Kubo, N. (1999) 'Development of scalable web crawler', *NEC Research and Development*, 40, 3: 334–340.

Terashima, N., Tiffin, J. and Rajasingham, L. (1999) 'Experiment on a virtual space distance education system using objects of cultural heritage', *Proceedings IEEE Computer Society Multimedia Systems*, 2: 153.

Tiffin, J. (1990a) 'Classrooms of the future – will there by any?', in B. Moon and C.M. Charles (eds), *Models of Teacher Education*, Townsville, Qld: James Cook University.

Tiffin, J. (1990b) 'Telecommunications and the trade in teaching', *ETTI*, 27, 3: 240–243.

Tiffin, J. (1994) 'Designing the virtual classroom', in G.E Bradley and H.W. Hendrick (eds), *Proceedings of the Fourth International Symposium on Human Factors in Organizational Design and Management – IV* (Stockholm).

Tiffin, J. (1998) 'The global virtual university: in search of the virtual class', *Proceedings 3rd International Open Learning Conference*, (Brisbane).

Tiffin, J. and Rajasingham, L. (1995) *In Search of the Virtual Class: Education in an Information Society*, London: Routledge.

Vygotsky, L.S. (1978) *Mind in Society: The Development of Higher Psychological Processes* (M. Cole, V. John-Steiner, S. Scribner and E. Souberman, eds), Cambridge, MA: Harvard University Press.

8

HYPERLEISURE

John Tiffin

Editors' introduction
The previous chapters describe utilitarian applications of HyperReality that are being
worked on in some form or another, even if it is only in the laboratory. This penul-
timate chapter is more speculative. Given that HyperReality is possible then what
would we like to do with it? It is an opportunity to rub the magic lamp of HR.
The field that seems most appropriate for this question is leisure activities and the
topic is broad enough to allow a general picture of life in an HR society. The closest
most of us get to rubbing a magic lamp is a punt on the pools or a lottery ticket.
If we win, all our dreams will be answered. Talking to a lottery researcher in
Germany a few years ago, John Tiffin asked if there was any research that looked
at the impact of a big win on the winners. Yes there was, but it was not some-
thing they wanted to broadcast. The bigger the win, the more likely the winner was
to commit suicide, become an alcoholic, get divorced or have a mental breakdown.
In consequence they had set up trauma teams to deal with winners as though they
had had a bad accident. HR has a similar dark side in that it seems capable of
catering to every self-indulgent dream. This chapter seeks to take account of this,
while it also addresses the new leisured class – the elderly – and the benefits HR
could bring to the future of ageing.

THE LITMUS OF LEISURE

A friend described how, at the height of the cold war, he was driving with
a colleague through the English fenlands when they heard the sound of sirens.
They had passed a military airbase where giant bombers were parked on alert
and their conversation had revolved around the precariousness of civilisation.

My friend pulled the car on to the grass verge and they got out and
leant on a gate that overlooked a field of grass that held a dazzling array
of spring flowers. They listened to the distant cadences of the World War
Two sound and his friend, although she knew he had stopped smoking,
offered him a cigarette. He took it and they lit up companionably. There
was no traffic, no one in sight, no radio in the car, no farm nearby and no

conceivable form of shelter from an atomic attack. In the silence after the sirens there was birdsong and the gentle rustle of vegetation in a soft breeze. Every sense seemed intensified. He felt heady from the combined aroma of grass, flowers, earth and tobacco. Every colour from every petal of every flower and every green from every blade of grass and leaf found every rod and cone in his retinas, and satisfied them. When they finished the cigarettes they climbed over the gate, lay down in the grass and for the first and only time made love to each other.

Estimates of the time between the sirens going off and the arrival of the first missiles varied between 8 minutes and half an hour. It gradually became clear to them that the sirens had some other purpose and the world as they knew it was not coming to an end. Their journey had a purpose and they had jobs to do. Although they were both married it was not to each other. And so they returned to getting on with their lives.

This is the litmus of leisure. The difference between the activities we find ourselves doing because we need to make a living, keep ourselves clean and tidy, rest and eat, protect and shelter ourselves and those we love, and the activities we choose to do when survival is not at issue.

Most of the chapters in this book describe ways in which HyperReality could make the activities we have to do for survival easier and more efficient. Will that mean more leisure in an information society? If so, for whom and what form will it take? This chapter looks at how HR introduces new sensory possibilities for leisure and the implications of this for self-realisation and ethical behaviour.

We may have leisure thrust upon us and some of us may even be born to it, but most of us, as Aristotle noted, achieve leisure through work. It can be at the micro-level of the breaks measured in minutes that we give ourselves as mini-rewards for doing parts of a task. Think of pausing for a cigarette, a coffee or a chat. At the mezzo-level of leisure we go to see a film or for a walk in the park or read a book; organised purposive activities that can be measured in hours. At the macro-level, industrial societies have organised the relationship between work and leisure to create the weekend, holidays and retirement where leisure is thought of in terms of days, weeks and years.

The leisure industry recognises that leisure activities are nested in other leisure activities. A film theatre will sell drinks, sweets and snacks. A resort hotel will provide swimming pools, aerobics classes, tennis courts, disco dancing and cultural events and in the hotel rooms there will be television, radio and a minibar. Theme parks create total artificial realities that envelop the visitor in a multiplicity of multilevel leisure activities.

The capability of coaction fields to coalesce in HyperReality allows for experiences to be embedded in experiences and so matches the needs of the leisure industry. A hotel complex adapted for HyperReality could have a HyperBar, a HyperTennis court, and HyperRooms. Anyone physically present in the real resort could have a drink with someone who was in another place, play tennis with a virtual tennis pro or dine with an escort of their choice selected from a set of options presented on a menu. It would in turn be possible to visit another HyperResort as a telepresence and enjoy its HyperReality amenities. Great hotels of the world could find additional income from teleguests who required no rooms or laundry and left no litter. A hotel chain could have the same sumptuous HyperShow running in all its hotels at the same time. Where a hotel today might be popular for its setting by a beach or among mountains, a HyperResort of the future might create its own thematic settings in VR and offer a weekend in Camelot or a safari in Jurassic Park or a trip down the Nile in the royal barge of Queen Hatsheput to see the fabled Land of Punt. Only the food, the drinks, the bed and the bathroom would need to be real and modern.

THE PURSUIT OF PLEASURE

Leisure is not simply the absence of an imperative to be involved in survival and maintenance activities. It is the active seeking of something for its own sake: the pursuit of pleasure, whatever an individual perceives that to be.

An empress of yore would summon slaves with a flick of a finger to carry out her desires. Today we surf the Web seeking our every information whim with a click of a mouse. HyperReality is, like the World Wide Web, a hypermedium. It promises the leisure-seeker the ability to surf from leisure activity to leisure activity. Intelligent leisure agents will, like the guides at one's elbow in a casbah, find whatever we want, but faster and with less risk.

Abraham Maslow wrote that we are a 'wanting animal' that 'rarely reaches a state of complete satisfaction except for a short time. As one desire is satisfied another pops up' (Maslow 1970: 24). With HR we have a technology that has the promise to instantly gratify our wishes, and even create new forms of esoteric pleasures for our jaded delectation (a holiday on Mars? an affair with King Charles (the second)? flight as a bird?). From the excesses of the Colosseum to child prostitution there has always been a dark side to the pursuit of pleasure when it is taken to extremes. Every new medium seems at some point to be appropriated for pornography. Today we worry about our children surfing the sleazy side of the Web. What will be the ethical issues arising from a medium like HyperReality that is so responsive to our desires? In mythology those who are given the gift of having their wishes granted are thereby destroyed.

OUTDOORS AS THE LOCUS OF LEISURE

In the episode at the beginning of this chapter appreciation of nature looms large as a form of leisure where the motivation is intrinsic. The need for beautiful scenery and natural surroundings is a major theme in the pursuit of leisure. Maslow, the taxonomist of human needs, believed people have aesthetic needs, but found this an area with which science is not comfortable (Maslow 1970: 51). Yet what else explains why city dwellers flock in their billions to places of outstanding natural beauty? The development of leisure time for much of the twentieth century was associated with the parks and recreation movement in industrial nations and seen as leading to a healthier and happier populace. However, in the declining decades of the century, the costs of mass tourism to the environment have become only too evident. In searching for beauty and solitude in the great outdoors we destroy it. From the time of Young's (1973) critique of 'blight-seeing', a growing body of literature has witnessed the erosive and polluting impact of tourism and there are increasing calls for sustainable levels of ecotourism (Krippendorf 1987; Pleumarom 1994; Boo 1990). HyperTourism could at least relieve some of the pressure on natural heritage sites by allowing people to visit places virtually without need of combustion engines. Would it be so very different from the way people tour in motor coaches? I have in mind being on a big powerboat as it cruised through the Great Barrier Reef of Australia and noting that the majority of the passengers were happily sitting inside watching a video of the boat they were on as it cruised through the Great Barrier Reef. They found the vicarious experience better than the real thing. HyperReality has the capability to facilitate a form of tourism in which the tourist need not be a part of the physical reality of the place they tour. The virtual tourist could visit one of the world's beauty spots for the afternoon and, immune to whatever disease, poverty, police, immigration laws or personal responsibilities might normally be attendant, be back in time for tea. Is this not simply an extension of what we do when we telephone someone in another country instead of actually going there to meet them? And if in the long run it conserves resources and reduces pollution by reducing the amount we travel, is that not a good thing? Or will we, like the people in the boat on the Barrier Reef, finish up doing both?

Where we can, we seek to create natural beauty or to link ourselves with the things of nature. Francis Bacon regarded gardening as 'the purest of human pleasures.' It has been a major leisure pastime since the days of the hanging gardens of Babylon. Pot plants along with canaries and cats are given time and space in even the meanest of slums, tenements and *favelas*.

HyperReality has the ability to respond to the way we like to domesticate and control nature. A city apartment could have a HyperGarden. Seeds could be selected from a catalogue of virtual plants that might grow, appear and behave like conventional plants, albeit with some differences. Kew Gardens could rent telepresences of its rare titan arum. The largest flower in the world, it naturally blooms on rare occasions in remote parts of the rainforests of Sumatra. The blossom is beautiful, but remarkable for the hideous stench it gives off. The patient observer could synchronously watch a virtual version of the Kew plant in the faint hope that something might happen and they would be among the first to see it. For the impatient, a virtual version of the flower could be created from the videos and photographs taken on the one occasion when the Kew plant actually did flower. This asynchronous virtual version of the titan arum could be programmed to flower on command and to emit a favourite fragrance.

Species of flowers, whose colours, dimensions, fragrances, and capabilities extend the very meaning of the concept of a flower, could be randomly generated from a vast set of possibilities lodged in a database for flowers. Instead of using crayons, children could design 3D flowers 'with silver bells and cockleshells' to bring home from school. A HyperGarden could grow and change while it was watched, perhaps like a human performer, in response to being watched, blossoming and waving its fronds to cries of delight from an audience or in graceful acceptance of a compliment. Plants could create olfactory symphonies and endlessly orchestrate their rearrangement. Virtual bird life of every kind and song and myth would be no threat to the vegetation, nor would any virtual animal chosen to inhabit a HyperGarden do any damage. Seasons could follow each other through the day or happen with a spoken command. There would be no need to weed or water or worry about the weather. Virtual pets, whether gryphons or herds of mouse-sized elephants, could wander at large in a HyperHome needing no house-training or food and having the gift to recognise when they were not wanted. It would be perfectly safe to have some splendidly coloured serpent hissing from a branch, because it would not be possible to eat any of the apples grown in the garden.

People pay more for rooms with a view but windows in industrial societies often give on to dreary street scenes. A HyperWindow in a HyperHome could have a superb view of Fujiyama on Fridays showing the mountain in all its moods as the weather and time of day changed. On Saturday the view could be of Mount Vesuvius and each day of the week could see the window with a different view from the series 'Famous Volcanoes of the World for your window on HyperReality'.

In 1996 Mount Ruapehu in New Zealand erupted. A site was set up on the World Wide Web that showed the image from a video camera set up to monitor the volcanic activity. Like the legendary university coffee site on the Web (a video shows the state of the coffee in the percolator in the staffroom) the purpose was initially utilitarian. Just as the coffee watchers could decide when best to take a coffee break, so New Zealand netizens could decide whether it was safe to go skiing on the slopes of the volcano. However, both web sites attracted worldwide interest and were visited by large numbers of people who found it interesting simply to look at the current state of a coffee pot and a volcano. The most popular site on the Internet at the time of writing this is known as 'Jennicam'. In this case a video camera continuously monitors a student's bedroom. Something most parents would prefer not to think about let alone look at is checked out on the Internet millions of times a day.

There is something window-like about the shape and function of a VDU linked to the Web and the idea of taking a break by glancing through it at the state of eruption of one of the world's finest and most dynamic volcanoes makes sense of the possibilities for HyperWindows. However, the idea of a window on the state of percolation in a pot of coffee is distinctly Zen and the global interest in the state of a student's bedroom suggests we are only at the beginning of understanding the voyeuristic possibilities of the views we could surround ourselves with in HyperReality (the state of the crowd in front of St Peter's Basilica? Timothy O'Leary's deathbed? the departure lounge at Narita Airport?).

There were in fact a number of video cameras placed at sites around Ruapehu for civil defence services to monitor the mountain. These were integrated with a steady stream of images from satellites and detailed maps of the area. This basic information could have been used in the manner described in Chapter 3 to create a three dimensional fly-over virtual reality of the volcano showing the state of its eruption in real time. This could have been made available in VRML on the Internet for access by any netizen anywhere anytime. In a HyperHome a selecting click on the view through a HyperWindow of a volcano could jack you into that volcano. From being at home thinking of going to see a volcano you could be flying over it like a bird, oblivious to the dangers of the molten volcanic bombs around you. Instant tourism to places that exist in real time, to re-creations of places that have existed and now reside in databases and to places that derive from the imagination becomes possible without any of the costs associated with moving people large distances and getting them to inaccessible places.

Associated with the need to get in touch with nature is the desire to feel fit and a craving for fresh air. Parks and recreation movements have encouraged outdoor fitness activities such as camping, climbing and cycling. The parks that serve as a city's lungs are full of skateboarders, strollers, joggers and people doing t'ai chi. Flying over the affluent suburbs of developed countries reveals a pattern of private pools. In the houses beside them will be a variety of exercise machines. Around the world people seek time each day in the places where they live for some form of exercise or relaxation whether it be yoga, aerobics or press-ups. Such activities are already conducted in primitive forms of HR. Skateboarders drift down sidewalks locked into the sensory world of their Walkmans and joggers pull out their mobile phones to conduct their affairs at the run. All these activities could be spliced with HR even if it was simply at the level of being able to consult maps, histories and biological data about bird life. However HR also opens possibilities of peopling a swimming pool with friendly dolphins or having a personal aerobics session from Jane Fonda.

Participation sport is linked to the fitness movement but has its own complex motivations that can vary from a need for self-actualisation to the need to dominate, compete or be part of a group with a strong sense of purpose. Communities respond to these needs and find space in recreation areas for sports. Where they do not, wasteland and streets become the venue for games, which may not always be socially acceptable. Running through our love of sport, fitness and the call of the wild is a predatory streak. In one form or another many people like leisure to hunt.

Pterodactyl, one of the first popular VR games, had two people in a strange multilevel place hunting each other with pistols while pterodactyls flew around ready to make a victim of the unwary. As any games parlour or video-game system shows, combat games are a popular leisure activity. David Goebels, the founder of Active Worlds, spoke of the possibilities of applying what he called 'extended reality' to the timber wolves that have been released into the Yellowstone National Park in the US. These animals are tagged to send signals to satellites so that their movements can be monitored. Goebels would like to see a micro video camera installed on their heads sending a continuous video image via satellite of what the wolves do. It gives new meaning to the idea of hunting.

Imagine a camera in the eye of a fish-hook linked by a small fibre line to the head-mounted display unit on a fisherman, then imagine that the fisherman is not even on the boat. He is in a fighting chair mounted in the safety of his study from where he can get on with his business until there is a strike when he switches realities to join the team on the boat and take charge of the robotised rod.

Sporting and recreation activities provide a clearly defined design challenge for VR developers. Interface devices between sportspeople and their sport can be created for simple cybernetic relationships. Shifting weight on a ski can affect what is seen on a screen to give the effect of skiing. Virtual reality was first developed to simulate flying. Devices for walking, cycling and driving in VR are well established. There is a race to develop games of virtual golf, tennis or cricket where a well-made intersection of a virtual ball with a real bat has a satisfying consequence. In HyperReality a golf game could be conducted on a real golf course in a coaction field between a real person with a real ball and a virtual person using a virtual ball. If St Andrews had a HyperWorld dimension it could multiply infinitely the number of people using its fairways without any damage to its famous turf.

INDOORS AS THE LOCUS OF LEISURE

As the incident at the beginning suggests, sex is one of the most basic leisure activities though normally, whether participatory or vicarious, it is an indoor pursuit and indoor pursuits exclude the natural world to create an artificial world in which we can live on the cusp of reality and fantasy.

Ernest Bormann's (1985) fantasy theory is an argument that we seek to understand ourselves and give meaning to our lives as we communicate within small groups by sharing stories. The metacommunication behind pictures in tourist brochures, posters and television commercials is that holiday makers will have a safe romantic adventure that could involve sex, the sensual pleasures of good food and drink, pampering by servants, dressing up and going to dinners, dances and theatre. The entertainment industry in theatre, cinemas, radio, but above all in television (Ibrahim 1991) seeks to vicariously satisfy our need to give meaning to life by helping us to make dramatic and cultural sense of events. In newspapers, magazines and books the publishing industry has a similar storytelling function. It is the integration of the stories and myths by which people live that takes place between the mass media, entertainment and tourist industries that constitutes what Max Horkheimer and Theodor Adorno (1993) called the culture industry, and the culture industry in all its manifestations is virtually the same as the leisure industry.

Chapter 2, this volume, drew attention to the use of the term *hyperreality* by Baudrillard (1988) and Eco (1986) for the way people sometimes seem to place greater credence on the mythological reality created by the media than on the physically real world in which they physically live. This *hyperreality* is made increasingly convincing by the growing intertextuality between the different components of the culture industry. Adverts for travel to London will have pictures of Big Ben, bobbies, buses and pubs and seek to evoke scenes from the works of Charles Dickens. Such stereotypes could form the paradigmatic elements of a London HyperWorld. The technology of HyperReality could mean that we become completely enveloped in the *hyperreality* of theme park versions of reality.

Harold Innis (1951) argued that the dominant medium of a society profoundly shapes the nature of that society. He demonstrated this by contrasting oral societies with literate societies. Marshall McLuhan took this thinking further, looking at the impact of television and information technology in the emergence of what he termed a global village (McLuhan and Fiore 1967). This captures the idea that a skein of electronic communications would allow the development of the local gossip of village communications on a global scale. The mass media have indeed become global in nature, but a major concern of many countries, first evidenced in the MacBride report (International Commission for the Study of Communication Problems 1980), is that mass media constitute not so much a market place for global chat as a global theatre for presentations, whose content is decided by a handful of international corporates while the rest of the world looks on as passive audience. With the exception of the broadcasting of some sporting and ceremonial events, mass media have become essentially asynchronous. There is a gap in time between the origination of mass media messages and their delivery to the mass audience. The intervening period allows for the gatekeeping function of editing. Critical theorists of the Frankfurt School have, since the 1930s, expressed concern that the culture industry through editorial control is a vehicle for imposing the hegemony of dominant ideologies. If HyperReality is a technology that enables us to live in *hyperreality*, whose *hyperreality* will it be? Whose stories will we identify with? Whose fantasies will we live?

An opposing view is held by adherents of the 'uses and gratifications' tradition of media studies first outlined in a book edited by Blumler and Katz in 1974. They see the content of mass media as something that is produced in response to the needs and desires of the masses. In a free society it is the users of the media that, by their collective choices, ultimately control the media. In this view any *hyperreality* created by the culture industry has been conjured up by the people themselves. Following this school of thought, HyperReality can be seen as a means for the public to construct popular fantasies.

The first mass-produced book was the Bible and there has always been a degree of divinity in the concept of authorship. Authors create the virtual world of their text. In this view readers seek to understand the author's meaning. In 1977, at the very time that Europe was being shocked by theological discourses that argued the demise of the deity, Roland Barthes shook the literary world by announcing the death of the author. This has become a key tenet of the postmodern movement. Once an author has created a text their role is over. Texts take on their own lives and can have as many interpretations as there are readers. It is the reader who creates a virtual world from the text and in so doing assumes the mantle of creator. From this perspective, virtual reality is the postmodern medium par excellence. The ingredients for a virtual world and the process for creating one can be downloaded in something more like six minutes than six days. From there on, the Adams and Eves of such worlds can take over genesis and make up their own mythology.

HR is a fully meshed medium like the telephone and, like the telephone, it will be a medium for chat and gossip. At the moment the Internet allows anyone anywhere with access to the Net to communicate with any other netizen by text, voice and video. HyperReality will bring to the Internet the added dimension of communicating through televirtual reality. It is the medium of a global village in which the virtual villagers can lean over their fence and, provided someone can throw in a camera, watch what is happening without being arrested or beaten up. Everyone in a global HyperVillage can have eyes and a voice. Potentially this is a medium of involvement, but equally it could provide a way to watch without involvement like a spectator at an execution.

The Internet is already seen by many as a medium of subversion, a Tower of Babel, a force for anarchy. The gatekeeper functions that control what is communicated by the mass media are not as yet in place on the Internet. Though governments and business seek to control the Internet in the way they control mass media today, this may not be possible to the same degree. It is possible that people could enjoy the freedom of access to HyperReality that has typified the telephone system in many countries for much of this century.

With HyperReality, virtual reality does not supercede physical reality. Teletravellers in HyperReality cannot escape the immutable laws of physics in the place where their bodies are, but they can sojourn as telepresences in another physical reality and interact with that reality without being subject to its physical laws. This sounds like the way gods and goddesses in Grecian mythology would on a whim come to live amongst mere mortals. They usually succeeded in fundamentally dislocating the rules and values by which ordinary mortals lived. The sex act described in the beginning of this chapter was outside the rules of behaviour for the people involved but they did it because they thought the rules no longer applied.

Sex will be possible in HyperReality between real individuals and virtual individuals, that is to say between atomically constituted people who are physically present in physical reality in one place and people who are virtual either in the sense that they are physically and synchronously present somewhere else or in the sense that they are activated by an artificial intelligence. In physical reality the real people would be wearing dildonic or sheath-like contraptions that were part of a comprehensive data-suit. The perception of the physically present person would, however, be that he or she was making love to another person. And indeed as far as the love was concerned they might well be. The telephone has long been a favourite device of lovers for expressing their deepest feelings. An advert by the Telecom Corporation of New Zealand advised lovelorn people to use the telephone 'to reach out and touch somebody'. With HyperReality this would not need to be metaphorical.

Of course a voice on a telephone sounds different to a real voice, but this is a question of bandwidth. If bandwidth becomes virtually limitless then we could imagine a day in the next century when it may be difficult to tell whether the person one is engaged in sexual relations with is real or virtual. Indeed, a virtual lover may be preferable. They would not snore and could always be improved on, made bigger or smaller in various ways, with more or less stamina, given the features of a favourite star and, if all else failed, they could be turned off.

With sex in HR there would be no transmission of bodily fluids. No one would get pregnant or contract Aids. There would be none of the consequences of conventional lovemaking, upon which we have built our moralities and laws, our family and social structures, and the beliefs and myths of who and what we are. Already the legal systems of countries that are becoming information societies seek to cope with new forms of marriage.

Medical advances mean that conception and pregnancy are no longer dependent on heterosexual coupling. HyperReality is part of a process that has run through the twentieth century of uncoupling sex from socially sanctioned breeding and its responsibilities.

The couple who thought they were faced with atomic annihilation sought in sex to find a meaning for their last moments. But when they realised that life was continuing as normal they never made love to each other again because they were locked into the social, legal and affective structures of their respective marriages. Yet, surely they remembered the moment in the flowers. In the virtual realities of our thoughts and dreams how many of us are faithful to our significant others? Who of us has not at some time substituted a virtual person for the one we held in our arms? In the privacy of our minds how many of us are monogamous? Sex has always been an activity in which we sought to reconcile dreams with reality. There comes now a technology that embraces the real and the virtual. In *The 120 days of Sodom* the Marquis de Sade (1987) imagines a castle at the end of the world where there are no limits to the excesses of human desire and the true nature of a humanity in which evil is an inextricable part can in consequence emerge. With HR there will be no need to go to the end of the world; anyone anywhere will be able to explore their limits. The extremes of depravity de Sade describes were in the virtual worlds of his books rather than in the physical reality of his everyday life and revisionist thinking sees him as part of the eighteenth-century movement of the Enlightenment, yet the influence of the virtual horrors he conjured up can be seen in the real world depravities perpetuated in the name of rationalism by twentieth-century nation states (Schaeffer 1999).

HOW MUCH LEISURE IN AN INFORMATION SOCIETY?

In the 1970s and '80s the information society was described as a Leisure Society (Bennington and White 1988; Kando 1980). It was assumed that automation and an infrastructure of information technology would mean less work. The growth of leisure time that characterised most of the twentieth century in industrial societies was expected to continue in the next century, as they became information societies. Machines would do the mundane tasks and the proportion of the population engaged in traditional forms of employment would decrease. There would be shorter working days in shorter working weeks for a shorter working life. The expansion of the leisure industry that we have seen in the last century would continue into the next where it would become a major part of the economy of the information society.

More recent views of the information society as a leisure society (Olszewska and Roberts 1989) see this concept in the same light as the paperless office:

a lovely idea that has yet to come. The societal restructuring that came with the application worldwide of neo-liberal policies, combined with the introduction of information technology in the 1980s and '90s, contributed to unemployment. Though this means an emerging class of people with time for leisure activities, it does not mean they have the financial, motivational or skill resources to take advantage of this. They live in cardboard cities not computopias. For those in regular paid employment the hours of work have tended to increase rather than decrease, suggesting that it is the nature of work rather than the amount of work that is changing. The electrification and automation of housework means less time spent on the chores of housekeeping. Day care and pre-school centres reduce the time involved in childcare, but the main effect has been to free women to take paid work. Modern transit systems have improved the speed at which they carry people over space but the consequent urbanisation has come to mean that many city workers spend a large part of their day travelling.

Another prediction for an information society is that it will be a telesociety. The development of an infrastructure of HyperReality has the potential to facilitate this. The vision of a lifestyle HyperCottage where teleworking, telelearning, teleshopping, telebanking and telemedicine reduce the time spent travelling seems attractive but the technology that makes it possible will also create a globally distributed workforce where going on strike is ineffective. Like the cottage industries that preceded the Industrial Revolution, the telecottage industries of the information revolution may prove to be places of hardship where whole families are involved for lengthy hours in the business of making a living.

Ibrahim (1991) finds that the working adult in an industrial society spends some eight hours asleep. Full-time workers on working days spend some eight hours at work. Much of the remaining eight hours goes on maintenance processes such as cooking, eating, hygiene, cleaning, caregiving, shopping and travelling. For the working adult there is not much time left for leisure activities except at weekends and holidays. The irony is that those in work who earn the money for leisure are those who have least time for leisure. Those who do have time for leisure in modern industrial societies are the young, the old and the unemployed – those with the least resources to enjoy it.

In developed countries the young have leisure. In early childhood the majority of waking hours are spent in play. At first the hours spent at school are few and many activities are more related to play than work. However, this gradually changes as the young person in an industrial society is prepared for working life and school takes on the mode of a work activity. Will the leisure years of childhood expand or diminish in an information society? There is a growing concern that learning in early childhood is a critical activity (Katz 1996). Around the world there are new initiatives with pre-schoolers (Boocock 1995). Kidnet becomes a global force and the

Internet a means for applying pressure to improve early childhood education (ERIC/EECE 1998).

The last century saw an increase in leisure time for the young in countries that legislated against child labour, but there is no reason to suppose that this trend will continue in an information society. As education adapts to the needs of an information society and becomes a lifelong process, young people find themselves under increasing pressure to continue studying after they start work. It is in the 'third age', the time between the end of work and death, that we can already see what promises to be one of the defining attributes of the information society: the emergence of the elderly as a new leisured class.

The legislation in industrial societies that institutionalised leisure time into working weeks and public holidays also entrenched the idea of retirement at a fixed age. At the beginning of the last century when life expectancy in industrial societies was below the age of retirement, schemes that encouraged people to save for their retirement years did well. Those who did get a golden watch tended to conveniently die shortly after. However, life expectancies in industrial societies have risen throughout the century. A person retiring today at age 65 can expect to live a further ten years or more and to be more active and healthy than was the case in the past.

The ageing of populations places strains on the ability of societies to support the elderly. Some countries look to the abolishment of institutionalised retirement and the reduction of state support for the elderly. Elderly people find they need to take part-time jobs to supplement pensions that have eroded in value. People with valuable skills and experience continue working on a part-time or contractual basis after they reach retiring age. Accrued knowledge and experience has value in knowledge work, people are healthier in their sixties than they were in the past and work in an information society is mental rather than physical.

The abrupt change from full-time work to full-time leisure that constituted retirement in industrial societies may be giving way to a gradual retirement process that begins at an earlier age with a steady increase of leisure time that matches the inclinations, aptitudes and ageing processes of an individual. There has been a trend in industrial countries over the last two decades for people to take early retirement from long-term jobs, then to work part-time with a greater concern for their lifestyle. The development of an infrastructure of HyperReality that facilitates telework could promote this process.

What is not clear is the extent to which life expectancy will continue to increase. One school of thought sees humans programmed by nature to die in their eighties. They point to the small number of people who live beyond 100 years. Another school sees no limits to the age people can reach with the development of genetic engineering and nanotechnology (Drexler 1990). Even if the increase in life expectancy does have limits, the decline of reproduction rates in developed countries means that they are faced with a progressively ageing population. The information society may well be a society of the elderly. Will HR be a boon to the aged in the way that spectacles have been or a bane to them like the plastic wrappings that old hands have such difficulty in dealing with?

THE ELDERLY IN HYPERREALITY

In writing this I have in front of me a vision of my father in his nineties waiting and wanting to die. His mind as keen as ever, a great cathedral of experience holding compassion, understanding and knowledge of a century, was slowly being bricked off from the rest of the world as his sensory apparatus decayed. He could not see to read, and how he loved a good book and the beauty of a rose. He could not hear the music that he had lived his life to. Speaking with him meant shouting. With his arthritic hands he could not write in his inimitable way or use the tools that had been an extension of himself. Walking in the park, where he knew every tree, was pain. He had all the time in the world for leisure but it availed him nothing. The irony for those who achieve the golden years of leisured retirement is that each year as they get older, their ability to partake of life declines.

Today's mass media are monaural and monocular and intended for normal hearing and sight. By contrast, the technology of HR can address the individual eye and ear and could in the future address each individual rod and cone of the fovea in each eye. With mass media the individual with poor eyesight and hearing has to buy glasses and hearing aids to compensate for their difficulties. With HR the medium can be tailored to the needs of the individual.

A HyperChair could be designed (perhaps as an adaptation of a motorised wheelchair) to allow a frail or disabled individual access to HyperReality. They could adjust volume without disturbing people in the same room, watch television with no difficulty, read a book without having to hold it or turn the pages, adjusting the size of letters to suit, and they could write and surf the Net by speaking into a small microphone. The person in the chair could control any electrical appliance in their environment and would be able to speak and interact with anyone else in the room as well as with any friends who wanted to televisit as telepresences. They too could televisit with old friends and relatives and lead a rich social life. Many of the HR technologies of leisure would be available to them. They could visit a HyperBar in a HyperResort and join a party in their honour at a HyperTable in a HyperRestaurant without ever having to leave their HyperChair.

Think further into the future when the number of elderly is greater, their life expectation is longer, they are healthier and they interface with HR through what they wear. With the kind of self-supporting suit that

Drexler (1990) describes, it becomes possible to think of the elderly having a new lease of active life. They could take part in HyperGames with the agility, strength, safety and support the suit gave them. Walking and general mobility would be easier. The suit could monitor their physical condition, advising if blood pressure was high or body temperature low, compensating by adjusting temperature and inclining for rest or providing a little more oxygen. And if necessary the suit could call in a virtual doctor and administer first aid.

My father's brother is also in his nineties but has escaped the worst ravages of age and can still read and type. The last time I visited him he asked me to explain what computers did. I retrieved my laptop from my car and showed him. An hour later he was excitedly surfing the Net. When I told him about HyperReality he said 'By God, I've got to live to try this.'

The traditional view of the elderly is one of progressive disengagement from their societal role (Teaff 1985) in preparation for death. Perhaps that made sense in times when physical strength was what counted for survival, but in an information society strength lies in wisdom, an elusive element most frequently found in the elderly. If the elderly are to take a major role in an information society, then HyperReality could have direct and major beneficial effects by extending their active sensory life to meet the challenges of what MacNeil and Teague (1987) call 'creative ageing'.

REFERENCES

Barthes, R. (1977) 'The death of the author', in S. Heath (ed. and trans.) *Image, Music, Text*, New York: Hill.

Baudrillard, J. (1988) *America*, New York: Verso.

Bennington, J. and White, J. (1988) *The Future of Leisure Services*, London: Longman.

Blumler, J.G. and Katz, E. (eds) (1974) *The Uses of Mass Communications*, London: Sage.

Boo, E. (1990) *Eco-Tourism: The Potential and Pitfalls*, US: World Wild Life Fund.

Boocock, S.S. (1995) 'Early childhood programs in other nations: goals and outcomes', *The Future of Children*, 5, 3: 94–114.

Bormann, E. (1985) 'Symbolic convergence theory: a communications formulation', *Journal of Communications*, 35: 128–138.

Drexler, E.K. (1990) *Engines of Creation*, London: Fourth Estate.

Eco, U. (1986) *Travels in Hyperreality*, London: Pan.

ERIC/EECE (Compiled) (1998) *A-Z The Early Childhood Educator's Guide to the Internet*, Champaign, IL: ERIC/EECE.

Horkheimer, M. and Adorno, T. (1993) *Dialectic of Enlightenment*, New York: Continuum. First published in 1947.

Ibrahim, H. (1991) *Leisure and Society*, Dubuque, IA: Wm. C. Brown.

Innis, H. (1951) *The Bias of Communication*, Toronto: University of Toronto Press.

International Commission for the Study of Communication Problems (1980) *Many Voices, One World: Communication and Society Today and Tomorrow: Towards a New*

More Just and More Efficient World Information and Communication Order, London: Kogan Page.

Kando, T.M. (1980) *Leisure and Popular Culture in Transition*, St Louis: The C.V. Mosby Co.

Katz, L.G. (1996) *Child Development, Knowledge and Teaching Your Children*, Champaign, IL: ERIC/EECE.

Krippendorf, J. (1987) *The Holidaymakers*, London: Heinemann.

McLuhan, M. and Fiore, Q. (1967) *The Medium is the Massage*, New York: Bantam.

MacNeil, R. and Teague, M. (1987) *Ageing and Leisure: Vitality in Later Life*, New Jersey: Prentice Hall.

Maslow, A.H. (1970) *Motivation and Personality*, New York: Harper and Row.

Olszewska, A. and Roberts, K. (1989) *Leisure and Lifestyle*, London: Sage.

Pleumarom, A. (1994) 'The political economy of tourism', *The Ecologist*, 244: 142–148.

Sade, Marquis de (1987) *The 120 days of Sodom and Other Writings*, London: Grove Press.

Schaeffer, N. (1999) *The Marquis de Sade: A Life*, London: Hamish Hamilton.

Teaff, J. (1985) *Leisure Services with the Elderly*, St Louis: Times Mirror/Mosby.

Young, G. (1973) *Tourism: Blessing or Blight?*, London: Penguin.

9

HYPERMILLENNIUM

John Tiffin and Nobuyoshi Terashima

Editors' introduction
The chapters in this book have explained what HyperReality is and speculated on its future impact in specific fields of activity. Bob Dylan sang, 'The times they are a-changing' and it is not just because of HyperReality. The connectivity between the factors of change in a networked global society means that HR will in its turn be impacted by such factors as the ageing of societies, the rise of feminism and the process of globalisation. In this concluding chapter the editors outline possible stages of development for HyperReality and place them in the broad context of the dynamics of the information revolution.

THE AGE OF HYPERREALITY

HyperReality waits in the wings. For HR to become the information infrastructure of the information society, we need a new generation of wearable personal computers with the processing power of today's mainframe and universal telecommunications where bandwidth is no longer a concern. Such conditions should obtain sometime in the first half of this century. Four stages of popular commercial development can be envisioned.

Stage 1. HR with PC and Internet (starting 2005)

HyperReality is currently being adapted at Waseda University for use on the Internet and could be available commercially in the first years of this new millennium. Special glasses will give the effect of looking through the VDU of a PC into a three-dimensional virtual reality and of being able to reach in with a gloved hand to interact with virtual people and objects in coaction fields selected from a menu. Virtual reality will be separated from physical reality by the bezel of the video display unit and the technology will not, therefore, be HyperReality in the full sense. It will, however, serve as a popular platform for the introduction of some aspects of HR and for trials of such applications as the HyperClass. It will be a tool for Internet users to develop their own coaction fields. This will be technologically at the low end of HR. It can be seen as part of the wider trend towards creating virtual worlds on the Internet, which has been led by developments such as Active Worlds and Black Sun and enabled by VRML (Virtual Reality Modulating Language).

Artificial Intelligence in this stage of HR would be in the form of the task-based agents, which are becoming common tools on the Internet. They would help to arrange and implement HR meetings and to facilitate the use of protocols and knowledge domains. Essentially they would communicate by text.

Stage 2. Flat-screen HR (starting 2010)

In the first decade of this millennium Digital High Definition Television (HDTV) will replace the kind of television we have known for the last forty years. Large flat-screen technology for television is at last becoming a commercial possibility. Telecommunication infrastructures of fibre-optic and coaxial cables will allow greater bandwidth. Computers and television sets are converging. Household computers will have enough processing power and bandwidth available to them for it to be feasible to think of using the same screen for TV and HR.

This stage of HR would be in line with the laboratory technology in ATR where HR was first developed by Nobuyoshi Terashima and his team. It will be possible for two or three people to sit or stand in front of a large screen, which could be curved. People using it would be less conscious of the edges of virtuality than they are with a PC monitor. Interacting aurally, visually and haptically with virtual objects and individuals would appear more seamless, provided the people using the system remain seated and oriented towards the screen. Research at ATR has shown that it is feasible for people from at least three different sites to work together in HR to solve complex problems with virtual models. The technology at this stage would be a high-end version of the stage 1 PC–Internet HR. Many of the examples given in this book are based on this format. It is 'from the waist up' HR and can be simplistically thought of as a three-dimensional form of videoconferencing. Avatars would be more realistic and identifiable and people will begin to take pains with their virtual appearance. Today people can make their own web pages or get them done professionally. In a similar way, in an HR world we can think of them turning to bespoke avatar tailors. Carl Loeffler, president of SIMATION, sees a conjunction of agents and avatars that would give AI a three-dimensional audio-visual presence. The person or creature or thing with which we interact in HR in ten years' time could be an artificial intelligence able to interact by talking, listening and moving.

Stage 3. Room-based HR (starting 2015)

By this time we can imagine the walls and ceilings in rooms being wired for information as today they are wired for electricity. As speech recognition and hand signals become the normal way to interface with computers, processing information could become the principal purpose of the partitions between rooms. Flat screen technology would be more pervasive and it is possible to think of walls, ceilings and even floors as surfaces for digital images. Rooms could become, on the lines of Plato's cave, artificial environments for the projection of sound and images and the interaction of real and virtual phenomena and of human and artificial intelligence. Initially one imagines such rooms being designed and developed for specialised purposes. Chapters in this book have been premised on the supposition that it would be possible to create HyperClasses, HyperSurgeries, HyperMuseums and HyperShops. Such ideas are feasible because the institutions involved, schools, hospitals, civic buildings and shopping malls, are likely to have the finances to invest in such developments and because they are already equipped for restricted domains of activity. In due course, however, HyperReality will become more generic in nature. The costs of using intelligent screen technology for walls and ceilings in a HyperRoom will decrease to a point where the technology will move into people's homes. In time it is possible to imagine a HyperKitchen, a HyperBathroom, a HyperLivingroom, a HyperNursery and HyperBedrooms in a HyperHome. In all of these HyperEnvironments it will be possible to introduce different kinds of coaction fields. As Chapter 2 explained, a person lying sick in a HyperBedroom could enter into a coaction field with their doctor in a HyperSurgery. Alternatively, the equivalent to reading in bed could be to enter into a coaction field in a HyperFantasy or an orchestral production from a HyperTheatre. A HyperKitchen may be a place to enter into a coaction field with a gourmet chef, but it will also be possible to check on the baby in the HyperNursery or a sick spouse in the HyperBedroom.

In these HR scenarios we can imagine a growing function for AI as expert systems within the knowledge domain of the coaction field. An AI nurse could monitor the patient in the HyperBedroom, an AI chef could help in the HyperKitchen and an AI nanny could tell fairy stories in the HyperNursery.

Stage 4. Universal HR (starting 2020)

As with radio and the telephone, the first use of HyperReality will be restricted to place. The Sony Walkman and the mobile phone made radio and the telephone portable and, in due course, HR will become portable and accessible anywhere. This could be as late as the third decade of the millennium. However, it could happen much sooner, due to such developments as high-flying aeroplanes acting as communications platforms for broadband telecommunications, and ideas for wearable computers. HR will then become a technology based on people rather than place. Initially we can imagine a laptop with an aerial giving HyperReality capability, but that still requires the user to look into a monitor. The difference will come with some combination of earphones, glasses and gloves that makes it possible to interrelate virtual reality and physical reality anywhere, in a manner that appears seamless. In this scenario people in HR will walk around talking to virtual people much as they do today with a mobile phone, except that they are seeing the person in full 3D as well as hearing them. Of course, a bystander would neither see nor hear the virtual person, unless they joined their coaction field.

In time we can expect the appearance of a totally enveloping smart suit and helmet containing a parallel processing network and a broadband transceiver. The human body will then have a layer of information technology separating it from the physical world. The effect would be to mediate all information to the human sensory apparatus. Lightwaves, soundwaves and tactile pressure from the immediate physical environment would be related to information sent by telecommunications. This integrated information would then be transmitted to the surface sensory apparatus of the body to give the effect of a seamless mixture of virtual and real phenomena. This HyperSuit would also operate in a reverse manner. It would sense the voice and the surface motion of its inmate's body and transmit voice and movement to any site with which the HyperSuit's inmate was in coaction. Following Drexler's idea that such a suit could be a flexible, self-supporting, self-articulating mould of its wearer (Drexler 1990), it would also be able to transmit sound and appearances to its immediate vicinity. As with the information the HyperSuit transmits by telecommunications, this may not necessarily be the sounds and appearance of the person within.

These scenarios suggest how HR could evolve as an infrastructure in a way that is consistent with existing trends. All the signs are that PC/Internet technology will continue to get less expensive to the user and consequently will become more universal. A billion users in the next decade is not impossible. Mobile phone usage around the world is on a similar track. Low Earth Orbiting Satellite Technology is at the launch stage. Television as we know it comes to an end in the first decade of this new millennium and will be replaced by Digital High Definition TV and with it bigger, longer screens that can read the language of computers. Architects design intelligent rooms in intelligent buildings for intelligent cities and Singapore has set itself the goal of becoming an intelligent country (one can only hope it starts a trend). The primary concern of such intelligent environments at the moment is with providing environmental control, security and computer and financial services. They are constructed to sense and respond to the movements and actions of people within them. They make widespread use of surveillance cameras and so provide the basic conditions for stage 3 HyperReality. In cities and buildings around the world cameras are increasingly being used for surveillance. As they become more ubiquitous, surveillance cameras become smaller, smarter and less obtrusive. Like electric lights and mobile phone transmitters, they are becoming an accepted infrastructural component of our surroundings. They go into place with little argument as to their implications and in so doing provide the basis for the stage 4 universal use of HR.

Once established, an infrastructure technology shapes the society that put it in place. Urban societies are defined by the existence of networks of roads, drains, water pipes, electric and telecommunication wires and cables. These create the conditions and enable the services by which people live, work and interact. We have grown accustomed to the idea of an imminent new telecommunications infrastructure. These information superhighways are to the information society what transport systems were to the industrial society, except that, instead of providing people with the capability of moving and processing vast quantities of atoms, they provide them with the ability to move and process vast quantities of information.

THE AGE OF THE INDIVIDUAL

The industrial society has been profoundly shaped by mass media, where the one to many mode of communication encourages a view of people as a mass audience. In contrast, HyperReality would appear to give primacy of communication to the self. As the intertwining of the real and the virtual becomes increasingly seamless, the Cartesian principle 'cogito ergo sum' is reasserted. The only reality in which a person can have any confidence is the knowledge of their own existence. If HR becomes the dominant medium of the information society, then it could facilitate the trend already in existence toward the third millennium as the age of the individual.

People have always modified their appearance to make an impression by the way they groom themselves, by the clothes and accessories they wear, by the way they speak and stand and gesture and the surroundings they place themselves in. However, physical reality imposes certain basic restrictions. In the real world people cannot change their age, nor is it a simple matter to change height, girth or sex. Yet people grow less inclined to accept what nature has endowed them with. They modify their bodies by plastic surgery, tattooing and piercing, seeking to reveal or conceal the real self (Neimark 1994). In HyperReality people can appear to be whatever they want to be.

Make-up takes on a new meaning when a person can study a simulacrum of themselves and transform it into however they want to appear for a meeting in HyperReality. They can subtly shift the contours of their face and body to seem younger or older or change their ethnic appearance or they can go to some gross extreme of fantasy. They can be a symbol, a talking tree, or a portrayal of their perfect self-image. How many of us could resist the childish wannabe within us? We could design different versions of ourselves for different occasions or go shopping for virtual personas like we go shopping for clothes. We could clone our perfect selves endlessly. And in doing this, to what extent is our sense of self then affected by the result?

The age of 'meism' came with the end of communism and its concern for community, and of socialism with its concern for society. Meism is fostered by privatisation and self-responsibility. Nation states and their concern for society are in retreat. They seek to reduce the taxation basis that redistributes wealth to provide essential services to all citizens. Instead the individual is held responsible for themselves. The user who pays calls the tune. National radio and television services no longer transmit common messages to all citizens on a limited number of government-regulated channels. Individuals can now flick through a multitude of deregulated channels in search of whatever satisfies them. In response, the media, increasingly global, panders to the pursuit of pleasure and entertainment. As the media move on to the Internet we can imagine intelligent agents scuttling through the recesses of the World Wide Web to garner the components of information cocktails mixed to satisfy our individual thirst for information or entertainment.

To click through a hundred radio or television channels or a dozen newspapers is to discover a dreadful sameness in an abundance of resources. Out of all the myriad events occurring around the world each day, only a handful are selected and presented as news. In the Age of the Individual I shall expect my intelligent newsagent to know that I am avid for any performance of Wagner by the Berlin Philharmonic and that I want to know at once whenever the local council discusses anything affecting my property. I want news, sports and events that interest me presented when I want, in the way I want it. Not the *Daily Chronicle* but 'The Daily Me' (Brand 1988).

The whole process whereby we receive mediated information about the world is being reversed. Instead of the editors of mass media systems selecting a restricted amount of information for mass consumption, the Internet has introduced the possibility for an individual to search for information that suits them. At the moment Internet users are in hunter mode. They use search systems to seek out the information they want. When intelligent agents do the searching for us, they will learn to cater to our tastes, our wants, our whims and our priorities, and they will get better and better at it. HR goes a stage further by locking us into coaction fields of predetermined knowledge domains. It surrounds us with the kind of knowledge and information that we want and shields us from whatever is discordant with our view of the world. In the age of 'I' will we narcissistically come to see a reflection only of that which we want to see?

147

THE AGE OF WOMEN

Feminist theory is associated with critical theory in arguing that the development of technology through the industrial society was not to the advantage of women (Grint and Gill 1995). This may not prove to be the case with a HyperReality infrastructure. Teleworking today allows people to work from home with a PC, a modem and a fax machine. The hours of individualised work can be adapted around the timetable of a baby, but there remains a problem in attending work meetings in synchronous mode. HyperReality would seem to be a solution to this. The primary thrust in its development has been to facilitate meetings.

This century has seen the ascendancy of women. The main thrust of feminism has been to redress the gender balance. Women have sought social and economic equality with men, but is there such a finely adjusted thing? Will the impetus of feminism push past the point of parity? Will the information society be one where women are dominant? Within Cyberfeminism there are calls for the restoration of matriarchy and for heterosexual men to accept female domination (Smith and Ferstman 1996).

The factors that enabled the rise of women in this century show no signs of abating. Women's control of their biology increases. Sex becomes separated from procreation. Husbands are no longer seen as essential. The use of sperm banks grows. Cloning becomes possible. The male role diminishes. There are more women than men and they live longer. Where universal democracy is the dominant form of government the numbers are with women.

Just as it is possible for someone using HyperReality to work from home, so too is it possible for someone travelling to use HyperReality to do a telecheck home. A woman on a business trip can, in her virtual persona, join her family for a meal or tell her children a bedtime story from her hotel. Whether from work or a supermarket, parents can look in on a nursery, respond to the questions of a babysitter and see for themselves that things are all right. If they are worried they can call on a doctor or a nurse to check on a child in HyperReality.

This is an extension of a process that began with the advent of the mobile phone. A person can be in many places at once and play multiple roles. Helen Fisher (1999) sees this in terms of evolution. For most of human existence, she argues, women stayed by the fire to look after children, prepare and preserve food and make clothes and they switched between these tasks. In contrast, men hunted and this is a more focused and continuous activity. In order to survive, women had to be good at multitasking and men at single tasking. The industrial society with its logical sequencing of tasks favours the step-thinking that men are good at. The Internet favors women's web-thinking and the way they can work in teams and find consensus (Spender 1995) and HyperReality is a supreme medium for meetings.

THE AGE WITHOUT AGES

Shakespeare described the acts of man in seven sequential stages from mewling infant to whining schoolboy, from the ages of active, mature life to the ages of the elderly. While there is some inexorability in the sequence of these steps in the process of ageing, in the information society we can overlap the activities represented in each age, so that they become more parallel than sequential.

Education is no longer something that happens between the ages of six and sixteen. It is becoming a lifelong activity. Learning begins at birth, if not earlier, and the greatest joy in learning may lie in later years.

The chapter on HyperLeisure questioned the abruptness of the retirement process in industrial societies whereby at the age of 60 or 65 a working person in full-time employment suddenly finds themselves with full-time leisure. Part-time work and short-term contracts are, in many countries, on the increase. Could we find our working lives becoming less demanding, being shared more with leisure and education activities, but extended further into the age that today constitutes retirement?

A major worry of governments in developed countries is the growing demand on national resources from the retired sector of the population. If the hours of work are reduced and if the nature of work is less physically arduous, need retirement come so early? Do children and adolescents need to be totally excluded from the work process? In our preoccupation with the abuse of child labour in developing countries, we forget the pride, the awareness, the learning and the growing up that take place when children are allowed to do meaningful work in the company of adults.

In HyperReality no one need know how old you really are.

THE AGE OF MOBILE CULTURES

Throughout history there have been great migrations and diasporas, but the sheer scale of movement of peoples in the twenty-first century has no equal. The aviation industry is preparing for a vast expansion of mobility in the new millennium. All the indications are that more people will travel further at faster speeds in an information society than was the case in the industrial society. To add to this we can also have instantaneous HyperTourism.

The transport industry, the tourist industry and the telecommunications industry intertwine. Seaports, airports and teleports integrate. The movement of bits of information by telecommunications is correlated with the movement of atoms in the cargoes of ships and people in aeroplanes. The universe of information and the universe of atoms are mobile in relation to each other. The implication of HR is that, wherever you are, you can always be somewhere else.

Physical mobility means that a substantial part of the world's population is transient in varying degrees and that there is a consequent intermixture of peoples from different cultures. Tourists go to places that have cultures different from their own in order to enjoy the difference. In response, indigenous cultures reinvent themselves in exaggerated ways. Immigrants bring their cultures with them and may form cultural enclaves or ghettos in their new countries, while maintaining links through modern transport and telecommunication systems with their homeland and their cultural roots.

One of the striking factors of our time is the persistence of cultures, despite the intermixture of populations and the globalisation of culture brought about by mass media. From the Second World War a main cause of wholesale armed violence has been some form of ethnic cleansing. Hitler used mass rallies and the new talking movies to create from Aryan mythology a culture that led to the Holocaust. Like film, HR provides a fertile place for the growth of myth and legend and, in addition, it is a medium for meetings and participation. It could serve as a vehicle for promoting racial solidarity and discrimination. Equally, however, it could, as much modern film and television does, provide a forum for intercultural issues that essentially seeks harmony.

When people use HyperReality for the thrills of being a HyperTourist and observing strange cultures in intimate detail, this could promote greater cultural understanding between peoples. On the other hand, the emphasis could be on what seems strange and bizarre. Will the HyperTourist see the other culture as a zoo? Will the people who are part of the indigenous culture view the HyperTourists as voyeurs? The migrant in a strange city could use HyperReality to stay in touch with their old country and culture. This might provide them with support while they seek to assimilate to a new country; on the other hand such links could tie them to their old culture in a way that inhibits assimilation.

New cultures develop on the Internet. Rheingold (1993) investigated the way people come together to create their own virtual communities. In such communities they can choose their sex and age, take on roles and behaviours and accept laws and codes of conduct that they find more acceptable than those that surround them in real reality. Such virtual communities will become stronger as they morph from their existence in text on the Internet, to existence in a HyperReality where the people they can see, hear, and touch, are like-minded.

People unhappy with the cultural roles into which they are born can seek to escape them for another country and culture. The history of migration shows that it is seldom a simple process and the consequences in the form of displaced people, ghettos and conflict between migrants and indigenous people are a major problem of our time. In HyperReality choosing a culture might seem as simple as clicking on a choice from a menu. Our avatars can have whatever ethnic characteristic we want to give them, but the mindset and meaning that goes with a culture cannot be acquired in the same way.

THE AGE OF REALISATION

Those of us who have enjoyed the benefits of living in an industrial society have only recently recognised that they carry consequences that will be inherited in the information society. We have polluted the world to a point at which we may have invoked inevitable and rapid climatic change. The third millennium will see violent storms, coastlands inundated by rising seas and agricultural production dislocated by rapidly changing weather patterns. In the first fifty years of the new millennium the population of the world will double, as will the problems we face if the pattern of behaviour that dominated the second millennium continues into the third. Will an infrastructure of HyperReality ameliorate or compound the consequences of the industrial epoch?

The infrastructure of automotive road transport of the last century has provided a killing field unmatched by the trenches of the First World War or Pol Pot. Yet it has also contributed in uncounted ways to an increase in human lifespan by the speed at which it becomes possible to get medical aid to the critically ill and injured. An infrastructure of HR will, like all new technologies, bring benefits and create problems, but in what field of human needs? Will it, like the technological infrastructure that supplies water, resolve a basic physiological need? Will it, like the infrastructure of electricity that lights our homes and streets, satisfy our primitive need to be free from fear of the unseen? Will it, as television and radio seek to do, cater to our need for belonging and love?

According to Maslow's (1970) theory of human motivation, these are the lower levels of a pyramidal hierarchy of human needs that have to be satisfied before we can turn to the higher levels of our needs for self-esteem and self-actualisation.

HR as an infrastructure would not appear to serve the lower levels of Maslow's needs hierarchy to any great extent. It is possible to exchange bits of information between the real and virtual dimensions of HR but not atoms. Although we speculate about HyperRestaurants and HyperBars, it is the social exchange in these activities that HR makes possible, not the actual transfer of food or drink. However, as discussed in Chapter 8, HR could be used to satisfy the physiological need for sex divorced from procreation.

The chapter on leisure also looks at a form of sanitised tourism with HR that protects people from the harsher realities of actually travelling to places where there are dangers. In the same way it could enable doctors to be telepresent at disaster sites and people to visit those with contagious diseases. HR could also be seen as a solution in cases where there were special security needs.

Satisfying the human need to belong and to love could be one of the strengths of HR. A fundamental characteristic of HyperReality is its ability to 'telephone' people together in 3D. It can bring loved ones close no matter how far apart they are, in a much fuller and more complete sense than is possible by telephone. They will be able to see, touch and hug and share activities together. HR in its own way breaks the bounds of time as well as space. Death need not part. With their virtual personas animated by artificial intelligence that has learned their habits and ways of responding and speaking, a significant other can in a sense live on long after the use-by date of their corporeal self. Costing little, needing no food or laundry, programmable and adjustable and with a 'save', 'close' and 'delete' function, they might even be preferable in this form.

'This above all,' Shakespeare's Polonius famously advised his son: 'to thine own self be true'. HR would seem to be the supreme technology for self-realisation. In HR we can seem to be anything we want to be and seem to do anything we want to do. Will we then, in using this medium in a search for ourselves, seek the wisdom or the foolishness that is within us?

In Maslow's theory, higher-level needs emerge as lower-level needs are satisfied, but what emerges when the highest level is satisfied and you can be whatever you want to be? According to Maslow, 'needs cease to take an active determining or organising role as soon as they are gratified' (1970: 57). It is like emptiness in a full bottle, 'a satisfied need is not a moti-vator' (ibid.). If HR is a supreme medium for becoming what one is capable of becoming, does this mean self-destructive self-gratification or will it allow us to transcend the needs of self and seek to satisfy the needs of humanity?

THE AGE OF MULTIPLE UNIVERSES

> No man is an Iland, intire of itselfe; every man is a peece of the Continent, a part of the maine; if a Clod bee washed away by the Sea, Europe is the lesse, as well as if a Promontorie were, as well as if a Manor of thy friends or of thine owne were; any mans death diminishes me, because I am involved in Mankinde. And therefore never send to know for whom the bell tolls; It tolls for thee.
>
> (*Devotions Upon Emergent Occassions*, John Donne)

John Donne writing in 1630 saw each person as part of a continent, and that continent as Europe. Until some 500 years ago, the world-view of literate and knowledgeable people was bounded by the continents they lived in. The age of discovery saw Europeans break out of their Mediterranean mindset. By the seventeenth century the basic outline of the continents and oceans was known and the globality of the world understood. Since then most people around the world have come to subscribe to a consensual view of physical reality that sees us as sharing space on planet earth.

Holland took a leading part in the exploration of the world. It was also famous for grinding and polishing glass. Dutch telescopes and microscopes enabled Galileo and Huygens to begin exploration of outer and inner space. This exploration continues, and the world's consensual view of physical reality now extends from the far galaxies of the cosmos to the components of an atom. In the process we have broken the code of human DNA, and discovered a scientific basis for Donne's belief that we are all related in some way.

The explorations that led to the scientific paradigm of the physical universe have been recorded in texts that have gradually been networked to form a virtual universe of our concept of the physical universe. Texts that purport to describe reality refer to other texts in the same field of knowledge. These texts in turn refer to other texts that refer to other texts in a vast self-supporting structure of taxonomically organised concepts, theory, logic and facts. Bibliographies, references and footnotes acknowledge these relationships and savants survive by citing them. The content of books about the physical world is nested in the content of other books about the physical world. New knowledge about the physical world is not acknowledged by the scientific community until it has been published, after a process of acceptance by readers and editors, and embedded through its references into the accepted body of literature on the subject. The knowledge of the physical universe in the great libraries of the world constitutes an integrated network of concordant ideas and facts that in sum constitute a virtual model of the physical universe.

Writers of fiction do not normally reference other works. However, careful reading of any text discloses the 'voices' of other texts (Bakhtin 1981), in the use of concepts, metaphors and myths. Bakhtin, echoing Donne, saw everyone and everything as interconnected by communication. Julia Kristeva (1969) coined the term 'intertextuality' to refer to the extent to which the meaning in any text is made up of the meaning in other texts. In S/Z Roland Barthes (1975) demonstrates how texts are woven from other texts and refer more to each other than to any physical referent. Richard Dawkins (1976), in introducing the concept of 'memes', suggests that there is something analogous to a genetic principle in the transference of ideas. In The Blind Watchmaker he further argues that memes can 'propagate from brain to brain, from brain to book, from book to brain, from brain to computer and from computer to computer' (Dawkins 1988: 158). In the process, memes can mutate and the consequent memic evolution is manifest in the evolution of cultures. The stories, songs, dances, carvings and paintings that once belonged in a unique way to individual cultures in separate parts of the world have, like the physical world, been 'discovered' and intertextualised in a global culture.

Intertextuality exists between as well as within different media, between as well as within the bodies of texts in different cultures and between as well as within different epistemes. Celtic ballads of a legendary King Arthur were incorporated into the written English texts of Sir Thomas Malory that were published by Caxton in 1485. These stories became elements of Ludovico Ariosto's romantic Italian mediaeval poem, *Orlando Furioso*, published in 1516. In the nineteenth century, Gustave Doré popularised Ariosto's stories for the mass reading public of industrial France by illustrating them with pictures of knights, fair maidens and dragons. These pictures have in recent times been a source of inspiration for films such as *Star Wars*. The mosaic of stories, characters and settings that arises from the memetic intertextual interlinking of sounds, images and words over time and between media has created a virtual universe of fiction that exists alongside the non-fiction virtual version of the actual physical universe.

In the nineteenth century fictional and non-fictional virtual universes were housed in books. The twentieth century has seen virtual universe databases expanded by film, video and audio tapes and digital discs. The digital foundations of the Web make it possible to intertextualise written text and spoken text, still images and moving images, film and television. The HyperText/HyperMedia mode that distinguishes the Internet and digital discs from the linear sequencing of mass media reinforces the intertextuality and referencing that holds virtual worlds of fact and fiction together. In the meshing of millions of computers, the Internet grows to resemble the neural activity of the brain. It enables a global virtual universe that we begin to call cyberspace. So far it has essentially been a medium for text and two-dimensional pictures. However, in its original conceptualisation (Gibson 1984), cyberspace was seen as a three-dimensional virtual reality universe of information. In the third millennium, with HyperReality, our virtual worlds of fact and fiction will become three-dimensional simulacra.

The process is already under way. A three-dimensional non-fiction virtual universe is coming into existence that seeks to model the physical world. Something like a remapping of outer and inner space is taking place. Geographers are reformulating the raw material that provides the basis of their 2D maps of the world into Geographical Information Systems (GIS). These allow 3D virtual realities of the world that show its shape, its continents and oceans, its rivers and mountains, and its cities. They can show the distribution of people and forests, the swirl of climate and weather and the flow of ocean currents. Simulacra of the world allow a person to take up a position in virtual space from where they can view the virtual earth and its virtual phenomena, then zoom in to a particular country, to a city, to a building, and finally into a room where they can meet with the avatar of someone who has clicked into a coaction field in HyperReality.

In this text HyperReality has primarily been described from the human perspective. We can imagine the development of virtual objects and life forms for a virtual planet earth that match our image of the physical world: the rocks, rivers and waterfalls, the fires and mists and the multitude of things made by people – motor cars and aeroplanes, footballs and skipping ropes, tennis rackets and canoes.

To begin with we may suppose that the correspondence between the virtual object and the real object would, like the correspondence between the real landscape and the virtual landscape, be essentially one of appearance. There is no need for a virtual engine to drive a virtual car, yet one cannot help but see the development of detail; the inner exploration; the creation of simulacra at the nanolevel. Virtual objects will acquire intrinsic structure, function and capability to match those of real objects. It will be possible to lift the hood on a virtual car and tinker with a virtual engine that appears to have the functionality of a real engine. A key characteristic of a coaction field is that it can be set to follow natural law.

This becomes particularly interesting in the development of virtual versions of living systems. Will it be possible to create a virtual person from a database of their DNA? The Human Genome Project to map the total DNA of the human species is largely completed. Atlases of the human body have been developed that make it possible to create a generic virtual person on the basis of the internal structures that compose an average human body. It then becomes possible to construct a virtual version of a person that includes their internal as well as external features and to observe the functioning of their organs at different levels. If we think in terms of a future where people wear an intelligent HyperSuit as a norm, then it would be possible to continuously measure and record blood pressure, temperature, pulse and respiration. The HyperSuit could have a homeostatic function, heating or chilling the body as necessary and advising its owner of health-threatening conditions. In emergency situations the HyperSuit could contact a doctor and initiate first-aid procedures. The HyperSuit could accumulate information over time so that virtual versions of a person scould be projected to show how they are likely to mature, age and have illnesses according to a calculus of their genes and style of living. It would also be possible to create an idealised virtual version of someone that, unlike their real self, could be reset and recycled to live again and again and again and, like the figures on Keats' Grecian urn, stay 'forever fair'.

And if it is possible to create a virtual version of a person, then it is possible to create virtual versions of forests, trees and leaves. We already create VR models of atomic structures, why not simulacra of stars and of nebula?

The age of discovery established the physical form of the world and its continents. In the age of reason that followed, Diderot and the *encyclopédistes* set out to create a comprehensive virtual world of knowledge that was rationally based on material facts. In so doing they shifted the authoritarian explanation of reality from theology to scientific empiricism. However, the accretion of knowledge of the physical world has so far been in alphanumeric form. The scientific approach has us measuring and enumerating reality and then interpreting and contextualising it in words. Encyclopedias provide a body of knowledge in the form of words and are organised alphabetically. As we succeed in mapping the surfaces of the virtual world in HR we will begin to populate it with virtual objects, virtual plants, virtual creatures and virtual people, created in detail on the basis of our understanding of how they are constituted atomically, the way they function and where they fit into the rest of the universe. This virtual world will have days and nights and seasons and climates and a history of events that parallel those of the real world. It will be part of a virtual solar system in a virtual universe.

John Tiffin remembers watching a child in the library of a primary school being restrained by teachers as he tried to tear up books that had hens, ducks and cows saying things to each other. 'Animals don't talk' the child screamed, with all the conviction of a six-year-old in outrage at a world that wilfully confused fantasy and fiction. Adults are saved from such experiences because, in their section of the library, virtual reality is clearly labelled as either 'fiction' or 'non-fiction'. Yet in such media as film and television fact and fantasy are blurred and Chapter 2, this volume, noted the coinage of the term 'hyperreality' by postmodernists to denote a faith in facts derived from fiction. We use fiction to explore the truth, facts are a favourite way to lie and the little boy in the library who saw the difference between fact and fiction was punished.

The virtual worlds of fact and fantasy created by texts (including film, video and web pages) are self-supporting in their referencing and intertextuality and have an existence that is independent of time and space and physical reality. The virtual worlds of HR are always by definition linked to physical contexts and real time. They overlap with the real universe. You cannot interact with a book or a film and you are not supposed to interact with web pages in a way that changes the text. However, you can interact with what is virtual in HR directly and dynamically so that it affects events in the places where people enter a coaction field.

HR intertwines physical reality with the virtual universes of fact and fiction in such a way that they can interact. Moreover, as the capability of coaction fields to combine increases, so too will the extent, comprehensiveness, detail and internal correspondence of the 3D sensory elements of its virtual universes of fact and fiction.

Limitations lie in the fact that coaction fields cannot combine when they have discordant characteristics (see Chapter 1, this volume), but this could serve to have the producers of coaction fields seeking the kind of congruence that intertextuality and referencing have achieved in virtual universes of text. Will we then be able to distinguish parallel universes of fact and fiction from reality itself, or does HR mean that they will merge in our minds as postmodernists such as Eco (1986), Baudrillard (1988), and Lyotard (1993) believe already happens? In inaugurating a coaction field, participants can establish and control the relationship between the virtual universes of fact and fiction and reality itself. A game of tennis in an HR coaction field can be matched to an actual game of tennis on an actual court. The laws of physics that determine the flight of a tennis ball can correspond in the virtual dimension to the real dimension. However, the players could decide to adjust the laws of gravity so that the tennis balls were slower or quicker and give themselves an audience of angels and demons. They can deliberately introduce elements of fantasy into their game. And one of the tennis players may be an expert system.

The crude reactive forms of artificial intelligence that we have today will begin to learn from experience. They will acquire the ability to reproduce themselves and will, unlike humans, do so each time at a higher level of capability and awareness. In the process they will acquire autonomy. Women may prove to be the dominant life form in the first years of the information society, but in the long term the information society may well be run by information creatures who will see us as we see chimpanzees: as their primitive progenitors.

Artificial life forms will have a lower-level needs hierarchy in order to survive, but can we imagine them with higher-level needs for belonging, love, self-esteem and self-actualisation? Isaac Asimov (1989) imagined a hierarchical set of rules that were by law built into intelligent robots. He was writing science fiction but raised the issue of an artificial intelligence having some set of priorities in its actions and whether this would be benign or malevolent for our species (Clarke 1994).

Is autonomous artificial intelligence with its own set of needs a serious possibility? The philosopher Hubert Dreyfus wrote *What computers can't do* in 1972 and then in 1993, as parallel processing became possible and seemed to herald something like neural networks in computers, *What computers still can't do.* These were powerful expressions of the limitations of AI when compared to human intelligence. But like most such critics of AI, Dreyfus adopts an anthropomorphic perspective. This derives from the direction given to early AI research by Alan Turing, one of the pioneers. He believed that human intelligence was computable and he provided a test for intelligence that has profoundly affected our view of intelligence. Quite simply, the Turing test states that if you cannot tell without looking whether the source of symbolic communications is a human or a machine, then you have artificial intelligence. In the absence of a clear, measurable definition, intelligence is judged to be something that is characteristic of humans. What if Dreyfus, in a similar vein had written 'What motor cars can't do' in 1900 and 'What motor cars still can't do' in 1920? He would have been able to show in both books that motor cars could not hop, skip or jump, climb ladders or walk downstairs and would never in a million years be able to dance the tango or play soccer. However, he would have to admit that, whereas in 1900 people walked in front of motor cars with flags, by 1920 motor cars could move at speeds of over 100 miles per hour and it is doubtful if, in the same time, humans had learned to run much faster. Wheels and legs are both means of locomotion but they are different and, because of this, they locomote differently. Nervous systems and computers are both means for processing information but they are different and, because of this, they process information in different ways. Computers cannot be intelligent like people, although we try to make them so, and they are good at mimicking, but neither can people be intelligent like computers, although a lot of them try, and some people seem to mimic computers.

We continue our voyages of discovery searching outer space for signs of intelligent life other than our own to add to our virtual universe of fact. (We already have alien life forms in our virtual universes of fiction that are so like distorted versions of ourselves as to suggest some Turing test at play.) We are more likely to find intelligent life coming towards us from the future out of the virtual universes of HyperReality that we created. Like us they will be able to interact in the coaction fields between the real and the virtual with us and with themselves. Just as we can have virtual avatar versions of ourselves that extend us in the virtual reality dimension of HR, these intelligent life forms will take shape in avatars and in robotic form that allows them to extend their activities into the physically real dimension of HR. This is when the great issues of HyperReality will emerge.

REFERENCES

Asimov, I. (1989) *Robot Dreams*, London: Victor Gollancz.

Bakhtin, M. (1981) *The Dialogic Imagination*, trans. C. Emerson and M. Holquist, Austin: Texas University Press.

Barthes, R. (1975) *S/Z*, trans. R. Miller, London: Jonathan Cape.

Baudrillard, J. (1988) *America*, London: Verso.

Brand, S. (1988) *The Media Lab: Inventing the Future at MIT*, New York: Penguin Books.

Clarke, R. (1994) 'Robot rules OK? An examination of Asimov's laws of robotics'. Published in two parts, *EEE Computer*, Dec. 1993 and Jan. 1994.

Dawkins, R. (1976) *The Selfish Gene*, Oxford: Oxford University Press.

Dawkins, R. (1988) *The Blind Watchmaker*, London: Penguin.

Drexler, E.K. (1990) *Engines of Creation*, New York: John Wiley.

Dreyfus, H.L. (1972) *What Computers Can't Do: A Critique of Artificial Reason*, New York: Harper and Row.

Dreyfus, H.L. (1993) *What Computers Still Can't Do: A Critique of Artificial Reason*, Cambridge MA: The MIT Press.

Eco U. (1986) *Travels in Hyperreality*, London: Pan.

Fisher, H. (1999) *The First Sex: The Natural Talents of Women and How They Are Changing the World*, New York: Random House.

Gibson, W. (1984) *Neuromancer*, New York: Ace Books.

Grint, K. and Gill, R. (1995) *The Gender Technology Relation: Contemporary Theory and Research*, London: Taylor and Francis.

Kristeva, J. (1969) *Semeiotike, Recherches pour une Semanalyse*, Paris: Editions du Seuil.

Lyotard, J.-F. (1993) *Toward the Postmodern*, Atlantic Highlands, NJ: Humanities Press.

Maslow, A.H. (1970) *Motivation and Personality*, New York: Harper and Row.

Neimark, J. (1994) 'Change of face . . . change of fate', *Psychology Today*, May–June.

Rheingold, H (1993) *The Virtual Community*, Reading, MA: Addison-Wesley.

Smith, J.C. and Ferstman, C. (1996) *The Castration of Oedipus: Feminism, Psychoanalysis and the Will to Power*, New York: New York University Press.

Spender, D. (1995) *Nattering on the Net: Women, Power and Cyberspace*, Ontario: Garamond Press.

INDEX

161

leisure industry 127–8; participation 30; protocols 23; translation and interpretation 102–3

collectionism 88, 89

Colombia, 'The Virtual University' 29

communication: coaction fields 9, 10; education 111–13, 122–3; humans and computers 86; humans and virtual characters 81–4, 99, 104; information technology and 2; Internet 100–2, 134–5; non-verbal 103–7; paradigms 33–6; syntagmatic dimensions 34; telephones 31–2, 33, 35–6; *see also* translation and interpretation

computer modelling *see* modelling

computer recognition *see* recognition

computer vision 20, 43, 44–7

control functions 11

control points 66–7, 69

coordinates: Sibson 66; texture fitting 62

creativity in computers 88–9

cues 45–6; *see also* non-verbal communication

cultures 150

CyberDance 75–6, 77

dance 75–6, 77

datasuits 31, 105–6, 135, 139–40, 145, 155

definition of HyperReality 1, 8, 41–2

deformation: animation 56, 74, 75; facial model 70; Free-Form models 58, 66–7, 69; virtual humans 58, 63–4

Degrees of Freedom (DOF) 63, 65

Delaunay triangulation 62

design 15–16

DFFD *see* Dirichlet Free-Form Deformations

Digital High Definition Television (HDTV) 143, 145

digital technology 33

Dirichlet Free-Form Deformations (DFFD) model 58, 66–7, 69

display of images 22–3

distal stimuli 31–2

Distributed Virtual Reality (DVR) 12, 13

DOF *see* Degrees of Freedom

domain knowledge (DK): coaction fields 5, 9, 10, 34; translation and interpretation 103

Dreyfus, Hubert 157–8

DVR *see* Distributed Virtual Reality

dynamic cues 46

Eco, U. 41, 133, 157

education 40–1; childhood 137–8; classes 111–13, 116–19, 120–1; communication system 111–13, 122–3; costs 123–4; HyperClass 17–19; industrial societies 110–11; The Virtual University 29

elderly people 138–40

electronic circuit evolution 95–6

ellipsoids *see* metaballs

emergence concept 88

emoticons 103

emotions, artificial 83–4

environment 129

Ethiopia 40

evolution: artificial brains 86–7; artificial life 85–96; electronic circuits 95–6; hardware 94–6; software 88, 90–4; virtual characters 81–2, 84–7

existence: virtual 39; *see also* life, artificial

expressions (facial) 72–3

faces: animation 70–4; modelling 57–8

fantasy theory 133

feature-based synthesis 48–50

features extraction 58, 72

feedback 9, 31

feminism 148

FFD *see* Free-Form Deformation

fitness activities 131–2

flat screen HyperReality 143–4

flowers 129–30

fly-overs 131

Free-Form Deformation (FFD) model 66–7

future 1, 142–58

games: coaction fields 157; virtual reality 132

gardening 129–30

Geographical Information Systems (GIS) 154–5

gestures 65, 106

GIS *see* Geographical Information Systems

glasses 6–7

globalisation: communication 100, 107–8; education 111, 115, 116; mass media 133

gloves 6, 31

hands: animation 64–9; modelling 58–9, 61–2; topography 67–8